FOUNDATIONS OF MODERN MATHEMATICS

THE PRINDLE, WEBER & SCHMIDT
COMPLEMENTARY SERIES IN MATHEMATICS

Under the consulting editorship of

HOWARD W. EVES

The University of Maine

Foundations of
Modern Mathematics

LYLE E. MEHLENBACHER

VOLUME SIX

PRINDLE, WEBER & SCHMIDT, INCORPORATED

Boston, Massachusetts   London   Sydney

Preface

The title *Foundations of Modern Mathematics* was selected from several possible titles, such as *Orientation to Modern Mathematics;* or *An Approach to Modern Mathematics.* All of these titles convey the idea that the primary objective of the book is to build the mathematical foundation needed for courses in modern mathematics. The names of these courses might include: linear algebra, modern algebra, foundations of geometry, analytic geometry, probability and statistical inference, vectors and matrices, calculus, and elementary function theory.

The term "modern mathematics" or "the new mathematics" has created a great deal of excitement during the past few years. This term is a big mystery to the average person, and especially to the parents of children who are learning modern mathematics. Further, the professional mathematician may wonder why we say modern mathematics, or new mathematics, when the material being taught is not new nor really very modern. Therefore it should be explained that the mathematics educator thinks of modern *courses* in mathematics instead of modern mathematics. These courses emphasize structure and basic principles along with techniques and appreciation for exact results, as opposed to the traditional courses in which emphasis was almost entirely upon techniques and algorithms.

The level of material in this book is elementary although we assume some previous knowledge of mathematical ideas and some skill in computation. The person who has taken, or has taught, the traditional courses in high school algebra and geometry should be able to follow this book. To profit fully from reading any book in mathematics, it is necessary to think and work, preferably with pencil and paper, while reading the book. With this cooperation the reader should find this book a useful bridge to further courses in modern mathematics.

The people who can profit from this book include:
(1) experienced teachers of mathematics at the elementary, junior

high, senior high, or junior college levels who must take a sequence of courses to prepare themselves to teach modern courses;

(2) twelfth grade students or college freshmen who need a bridge course before they go on to the courses required for a major in mathematics;

(3) students who need some knowledge of mathematics for courses in business, economics, biological science, or liberal arts;

(4) parents who want to learn something about "the new math" and are willing to spend some time and effort to learn it.

This book has been used in several preliminary forms as the text material in classes for high school teachers. This course is the first in a sequence of ten, each carrying three semester hours of credit. The sequence continues with courses in linear algebra, modern algebra, probability and statistical inference, modern concepts of calculus, fundamental concepts of geometry, theory of functions, and optional courses in the theory of numbers, analysis, and numerical analysis. There is sufficient material for a three semester-hour course.

LYLE E. MEHLENBACHER

Contents

CHAPTER ONE. THE MATHEMATICS LANGUAGE

CHAPTER TWO. THE CALCULUS OF STATEMENTS

CHAPTER THREE. UNIVERSES, SETS, BOOLEAN ALGEBRA

CHAPTER SIX. GEOMETRIES, PURE AND SIMPLE

CHAPTER SEVEN. ANALYTIC GEOMETRY IN THE
CARTESIAN PLANE

CHAPTER EIGHT.   ANALYTIC GEOMETRY IN 3-SPACE

CHAPTER NINE.   THE CIRCULAR FUNCTIONS

CHAPTER ONE

The Mathematics Language

1.1. THE MATHEMATICS LANGUAGE. Just as in beginning any other language, we start the study of mathematics with a basic vocabulary containing those words that are essential to the construction of statements or sentences. The words are placed in a sentence in accordance with certain well-defined rules of grammar so that the sentence will state a fact, express an idea, or formulate a question that will challenge the mind of the reader. As the student of the language gains experience in making statements in the new language, the vocabulary must be extended by the addition of new words, and additional grammatical rules must be evoked.

Mathematics is an exact language, which requires simple but well-defined words and strict compliance with grammatical rules. A sentence in mathematics must convey an exact message to the reader. Sentences with shades of meaning or sentences that permit more than one interpretation cannot be tolerated in this language. The writer of a sentence in mathematics must know what he wants to say, and he must be sure that his sentence states his message.

As a general rule, the words in mathematics are so short and simple that we usually call them *symbols*. A symbol in mathematics may require several words in another language to express its equivalent meaning. Consequently, a sentence in mathematics may require relatively little space compared with the space needed to say the same thing in another language.

The basic vocabulary in mathematics consists of nouns, verbs, connectives, and auxiliary words. The grammatical rules regulating the construction of sentences in mathematics are the same as the grammatical rules used in English.

1.2. NOUNS.   The first nouns in our basic vocabulary are numbers, which will be represented by the usual symbols, such as

$$1, 2, 3, 3.14, 1/2, 1/4, 1000, \sqrt{5}, 3 + 4, 3^2,$$

and many, many more.

These symbols are not unique. In fact, the same number may be written in many ways, for example,

$$5 = 2 + 3 = \sqrt{25} = 7 - 2 = 1 + 2 + 2 = 3 - 1 + 4 - 2 + 1, \text{ etc.}$$

The symbols for numbers are called *numerals*. Fortunately, the same symbols, or numerals, are universal in the modern world. However, the spoken names of the numbers are different in different parts of the world.

It is a common practice in mathematical language to make use of letters of the alphabet, or of the Greek alphabet, or other symbols, to stand for numbers, or numerals. Such symbols are called *pronumerals*, or *variables*. We prefer to use the term *variable*. We assume that the reader is familiar with the numerals and variables commonly used in ordinary arithmetic and algebra.

MATHEMATICS VERBS	ENGLISH EQUIVALENTS
$=$	Equals, is equal to, is the same as
$<$	Is less than, precedes, is to the left of
$>$	Is greater than, follows, is to the right of
$\parallel$	Is parallel to
$\exists$	There exists, there is, there are. This symbol is called the "existential quantifier."
$\in$	Belongs to, is an element of, is a member of
$\rightarrow$	Implies. "$p \rightarrow q$" is read "p implies q," or, "if p, then q," or, "p is a sufficient condition for q."

1.3. VERBS. The verbs in our basic vocabulary, with their equivalents in the English language, are listed below. We should note that, even though the word may have several equivalents in English, its meaning is quite definite.

In mathematics, new verb symbols may be formed by reversing verb symbols, by combining two symbols into one, or by negating verb symbols. A verb symbol may be negated by writing a vertical line or a slash through it. New verb symbols obtained from those in the first list are included in the following table.

MATHEMATICS VERBS	ENGLISH EQUIVALENTS
$\leftarrow$	Is implied by. "$p \leftarrow q$" may be read, "p is implied by q," or "if q, then p," or "p is a necessary condition for q."
$\leftrightarrow$	Is equivalent to. "$p \leftrightarrow q$" may be read, "p is equivalent to q," or, "p is a necessary and sufficient condition for q," or "p, if and only if q," or "iff p, then q."
$\nrightarrow$	Does not imply
$\nleftarrow$	Is not implied by
$\nleftrightarrow$	Is not equivalent to
$\neq$	Is not equal to, is not the same as
$\nexists$	Does not exist, there is not, there are not
$\notin$	Does not belong to, is not a member of, is not an element of
$\leq, \ngtr$	Is less than or equal to, is not greater than. These symbols have the same meaning.
$\geq, \nless$	Is greater than or equal to, is not less than. These symbols have the same meaning.

1.4. CONNECTIVES. A connective, or conjunction, is used to connect words, phrases, clauses, or sentences. We need three connectives in our basic vocabulary.

MATHEMATICS CONNECTIVES	ENGLISH EQUIVALENTS
$\cup$, $\vee$	Or, the union of. "$A \cup B$ is read "A or B," or "the union of A with B," or "the disjunction of A with B." These are the disjunction symbols, and the first symbol is called "cup."
$\cap$, $\wedge$	And, the intersection of, the conjunction of. "$A \cap B$" is read "A and B," or "the intersection of A and B," or "the conjunction of A and B." These are the conjunction symbols, and the first symbol is called "cap."
$\sim$	Not. "$\sim p$" is read "not p." This is the negation connective.

We shall use the cup and cap symbols throughout this book. The second connective symbol shown in the table above is sometimes preferred for use in the calculus of statements in the environment of Chapter Two, while the cup and cap symbols are reserved for the union and intersection of sets in Chapter Three.

The tilde symbol, $\sim$, will appear again in Section 3.12 as an equivalence relation. It is not uncommon in the language of mathematics to assign certain symbols quite different meanings, depending sometimes upon the placement of the symbol in relation to other symbols and sometimes upon the environment of the discussion. This same symbol appears again in geometry as a verb symbol meaning "is equivalent to," and as part of the symbol, $\cong$, meaning "is congruent to." In each case, the symbol is an equivalence relation between triangles or other polygons.

1.5. AUXILIARY WORDS.   Words that cannot be classified as nouns, verbs, or connectives are called *auxiliary words*. These auxiliary words or symbols are necessary parts of the language. The following table contains some of the more important auxiliary symbols for mathematics.

AUXILIARY SYMBOLS	ENGLISH EQUIVALENTS
$\ni$	Such that, so that
$\forall$	For every, for each, for all. This symbol is usually placed at the beginning of a statement. It is called the "universal quantifier."
$\cdots$	Whenever certain numbers may be left out without loss of meaning we use three dots to take the places of the numbers not written. For example, we write "1, 2, 3, $\cdots$, 100" to mean all of the positive integers from 1 through 100.
(parentheses), [brackets], {braces}, vinculum	These are symbols of grouping used for the purpose of indicating a combination of numbers which is to be considered as a single number.
$+, \times, -, \div,$ $\sqrt{\ }$, etc.	The usual symbols for combining numbers, binary and unary operations, will be included in the same sense as in ordinary arithmetic.
Exponents, coefficients, decimals, radicals, fractions, etc.	All of the ordinary symbols and devices in common usage in elementary arithmetic and algebra will be employed in the usual sense.

1.6. SENTENCES IN MATHEMATICS. A sentence in mathematics must make strict use of the rules of grammar in order to ensure that it says exactly what it is intended to say. One of the most common errors in writing mathematical sentences is the omission of punctuation. A mathematical sentence requires punctuation and should end with a period in the same way as a sentence in English.

A sentence in mathematics may be simple or compound in the same way as a sentence in English.

A sentence in mathematics may be correct grammatically and it may express exactly what it is intended to say without regard to whether it is a true statement or a false statement. In mathematics some statements are true, some are false, and some statements are conditionally true or conditionally false. In general, if a mathematics sentence contains only numerals for its nouns, it must be either true or false. If the mathematics sentence contains variables for its nouns it may be true only for certain numbers in the places of the variables. Such a sentence is called an *open* statement. Those numbers which, when they are put in the place of the variables in an open statement, make the statement true, belong to the *solution set*, or the *truth set* for that open statement.

An open statement in which the verb is = is usually called an *equation*. A statement in which the verb is one of the "inequality verbs," $<, >, \leq, \geq$, is called an *inequation* or an *inequality*. An open statement in which the truth set includes every real number is called an *identity*.

The following are examples of sentences in mathematics language:

a. $2 + 3 = 5$.

b. $2x - 3 = 7$.

c. $2x + 5x = 7x$.

d. $\sqrt{25} = 7$.

e. $3x + 4y = 9$.

f. $(x < y) \cup (x > y)$.

g. $(2 < 3) \cap (1/2 > 1/3)$.

h. $[(4 + 6) > (5 + 3)] \cup [(5 + 4) < (3 + 2)]$.

i. $(x < y) \rightarrow \exists t > 0 \ni x + t = y$.

1.7. PROPERTIES OF EQUIVALENCE VERBS. The verb symbol = is a member of a class of symbols that are called *equivalence verbs*, or *equivalence relations*. Other members of this class are $\sim$, meaning "is similar to" in reference to geometric figures, and $\|$, meaning "is parallel to" in reference to lines in a plane, and many others. We shall use the symbol r to represent any equivalence relation. Then, an equivalence relation r is required to have the following properties:

1.7.1. *Reflexive Property.* $\forall a, a \, r \, a$.

1.7.2. *Symmetric Property.* If $a \, r \, b$, then $b \, r \, a$.

1.7.3. *Transitive Property.* If $a \, r \, b$ and $b \, r \, c$, then $a \, r \, c$.

1.8. PROPERTIES OF INEQUIVALENCE VERBS. The four verbs $<, >, \leq, \geq$, are called *inequivalence verbs*. When we test them for the properties of equivalence verbs, we find that only the verbs $\leq$ and $\geq$ are reflexive, none of these verbs is symmetric, but all four of them are transitive. These four symbols consist of two pairs of symbols with their reverse symbols. We may characterize all four of the inequivalence verbs by using the representative symbol **R**, and its reverse symbol **Я**. For example, if **R** stands for $<$, then **Я** will stand for $>$. The following five statements characterize the four inequivalence verbs:

1.8.1. *Transitive Property.* If a **R** b and b **R** c, then a **R** c.

1.8.2. If a **R** b and c is any number, and if $+$ denotes ordinary addition, then $(a + c)$ **R** $(b + c)$.

1.8.3. If a **R** b and $c > 0$, and if $\times$ denotes ordinary multiplication, then $(a \times c)$ **R** $(b \times c)$.

1.8.4. If a **R** b and $c < 0$, and if $\times$ denotes ordinary multiplication, then $(a \times c)$ **Я** $(b \times c)$.

1.8.5. If $-b$ denotes the additive inverse of b, such that $b + (-b) = 0$, then a **R** b if and only if $a + (-b)$ **R** 0.

1.9. THE CONNECTIVE OF NEGATION, $\sim$. The mathematical symbol $\sim$ negates a statement when it is placed in front of the statement. Thus, the statement p is negated by writing $\sim p$. If p is a true statement, then $\sim p$ is false, and conversely. In the mathematical sense there is no ambiguity about a negated statement.

In English, most sentences can be negated simply by placing the word "not" at the proper place in the sentence. However, certain sentences in English are difficult to negate. It is frequently difficult to form proper negations of sentences that contain the word "all" or "some." For example, the negation of the statement "all prime numbers are odd" may have the forms:

a. All prime numbers are not odd.
b. Not all prime numbers are odd.
c. No prime number is odd.
d. Some prime numbers are even.
e. Some prime numbers are not odd.
f. At least one prime number is even.

Since the statement in the example is false, then its negation must be true. Statements b, d, e, and f are true while statements a and c are false. It follows that statements b, d, e, and f are proper negations of the original statement.

PROBLEM SET 1.1

1. Find the truth set for each of the following open statements:
   a. $6x + 2 = 26$.
   b. $x^2 + 6 = -5x$.
   c. $3y + 1 > 6$.
   d. $x^2 \leq 25$.
   e. $(x + 1)^2 - 1 = x^2$.
   f. $2x + 3 \not< 7$.
   g. $x + y = 5$.
   h. $2x - y \not> 3$.

2. Write the English translation for each mathematics sentence:
   a. $\forall x \cap y, \exists z \ni x + y = z$.
   b. $\forall n, \exists k \ni (2n + 1)^2 = 8k + 1$.
   c. $\forall e > 0 \exists d > 0$
      $$\ni \{[-d < (x - a) < d] \rightarrow [-e < f(x) - b < e]\}.$$

3. Write the mathematics translation for each English sentence:
   a. If the triple of a certain number is increased by five the result is thirteen.
   b. Every positive integer has an immediate successor which is one more than itself.
   c. If a number is a real number, then its square is positive.
   d. There is a positive integer whose square is 14641.

4. Invent a symbol for each of the following verbs and test this verb for the reflexive, symmetric, and transitive properties:
   a. is similar to;
   b. is congruent to;
   c. is the spouse of;
   d. disagrees with;
   e. is parallel to;
   f. was born in the same town as.

5. Replace the symbol **R** by the correct inequivalence symbol:
   a. $(x - 5) \mathbf{R} (x - 9)$.
   b. $(2x + 1) \mathbf{R} (2x - 3)$.
   c. $47/49 \mathbf{R} 41/43$.
   d. $577/408 \mathbf{R} \sqrt{2}$.
   e. $(x^2 + 1) \mathbf{R} 2x$.
   f. $1.732 \mathbf{R} \sqrt{3}$.

6. Write two or three correct negations for each of the following statements:
   a. Detroit is north of Canada.
   b. All good automobiles are made in or near Detroit.

c. Joe is always away from home.

d. If a is parallel to b, and b is parallel to c, then a is parallel to c.

e. All women who are not mortals are witches.

f. Some students take courses in mathematics and some students have a good time.

g. All honest men are fools.

h. Some positive integers are divisible by three if the sum of their digits is divisible by three.

7. Complete the following statement to make a true statement: "If both $a \leq b$ and $a \geq b$, then . . . ."

8. Let p mean "Joe is a democrat," q mean "Joe received more votes for dog catcher than Jack received," and r mean "Joe was elected to the position of dog catcher." Write the English translation for each of the following and discuss the possibilities for each statement to be true or false:

a. $p \cup q$.

b. $p \cap q$.

c. $p \cap r$.

d. $q \cap r$.

e. $p \rightarrow r$.

f. $r \leftrightarrow p$.

g. $(p \cap q) \cup r$.

h. $(p \cap q) \rightarrow q$.

CHAPTER TWO

The Calculus of Statements

2.1. TRUTH VALUES.   If it is possible to assign a "truth" value to a statement, then the statement is called a *proposition*. If a proposition is true it is assigned the truth value T. If a proposition is false it is assigned the truth value F. Each proposition must have the truth value T or the truth value F. No proposition can have both truth values T and F. No proposition in our discussion can have neither truth value.

We are not concerned about the meaning of truth in any philosophical or ordinary sense. Just how a proposition may obtain the truth value T or F is of no concern. We shall agree to accept the truth values of propositions as they are assigned or derived. If the truth value of a proposition agrees with our usual belief, we feel pleased. If not, we shall accept it without complaint.

Propositions will be designated by lower case letters from the middle of the alphabet, p, q, r, ... New propositions may be formed by combining propositions through the use of connectives, certain verb symbols, and negations. The truth value of a combination of propositions may be calculated by means of a device called a *truth table*.

2.2. THE NEGATION TRUTH TABLE.   The truth value T is opposite to the truth value F. If the proposition p has the truth value

p	$\sim p$
T	F
F	T

The Negation Truth Table

T, its negation, $\sim p$, has the truth value F. If p has the truth value F, then $\sim p$ has the truth value T. These facts are given in the *negation truth table*.

2.3. THE CONJUNCTION TRUTH TABLE.   Given two propositions p and q, the compound statement $p \cap q$ may be read "p and q," or, "the conjunction of p and q," or, "the intersection of p and q." The truth value of $p \cap q$ is T if and only if both p and q have the truth value T. If the truth value of either p or q, or of both p and q, is F, then the truth value of $p \cap q$ is F. The *conjunction truth table* contains the truth values for each of the four combinations of truth values of p and q.

p	q	$p \cap q$
T	T	T
T	F	F
F	T	F
F	F	F

The Conjunction Truth Table

2.4. THE DISJUNCTION TRUTH TABLE.   The compound statement $p \cup q$ is read "p or q," or, "the union of p with q," or, "the disjunction of p with q." The "cup" symbol will designate the *inclusive disjunction*. The inclusive disjunction, $p \cup q$, has the truth value T, if either p or q has the truth value T and the other has the truth value F, or if both p and q have the truth value T. The inclusive disjunction, $p \cup q$, has the truth value F if and only if both p and q have the truth value F.

When we use the connective symbol $\underline{\cup}$, the "cup with a saucer," we obtain the rarely used *exclusive disjunction*. The compound statement $p \underline{\cup} q$ has the truth value T if and only if either of p or q has the truth value T and the other has the truth value F. The exclusive disjunction $p \underline{\cup} q$ has the truth value F if and only if p and q have the same truth values, either both have the truth value T, or both have the truth value F.

The *disjunction truth table* contains the truth values for both the inclusive disjunction and the exclusive disjunction for each of the four combinations of truth values of p and q.

p	q	$p \cup q$	$p \cup\!\!\!\!\_\ q$
T	T	T	F
T	F	T	T
F	T	T	T
F	F	F	F

The Disjunction Truth Table

2.5. EQUIVALENT COMPOUND STATEMENTS. If two compound statements have the same truth values in the same row of a truth table, or, in other words, the columns of their truth values in a truth table are identical, then the two compound statements are said to be *equivalent*. We shall use the verb symbol = to mean "is equivalent to" in the sense just defined.

For an example, we calculate in a truth table that the compound statements "$p \cup q$" and "$\sim(\sim p \cap \sim q)$" are equivalent, or we show that

$$p \cup q = \sim(\sim p \cap \sim q).$$

p	q	$\sim p$	$\sim q$	$\sim p \cap \sim q$	$\sim(\sim p \cap \sim q)$	$p \cup q$
T	T	F	F	F	T	T
T	F	F	T	F	T	T
F	T	T	F	F	T	T
F	F	T	T	T	F	F

2.6. THE LAWS FOR COMPOUND STATEMENTS. Three important laws of combinations apply to compound statements in which conjunctions and disjunctions occur. These laws will appear whenever we employ two or more elements in combination. The laws are:

2.6.1. *The Commutative Laws.* $p \cap q = q \cap p$, and $p \cup q = q \cup p$.

2.6.2. *The Associative Laws.* $p \cap (q \cap r) = (p \cap q) \cap r$, and $p \cup (q \cup r) = (p \cup q) \cup r$.

2.6.3. *The Distributive Laws.* $(p \cap q) \cup (p \cap r) = p \cap (q \cup r)$, and $(p \cup q) \cap (p \cup r) = p \cup (q \cap r)$.

The commutative laws for both conjunction and disjunction are verified simply by inspection of the conjunction and disjunction truth tables, respectively. The associative laws and the distributive laws are verified in a truth table for three propositions p, q, r, containing eight rows for all possible combinations of truth values for three propositions. In the following truth table we write the combinations of propositions vertically in order to conserve horizontal space. We number the columns for easy reference.

1	2	3	4	5	6	7	8	9	10	11
p	q	r	$p \cap q$	$q \cap r$	$p \cap r$	$q \cup r$	$(p \cap q) \cap r$	$p \cap (q \cap r)$	$(p \cap q) \cup (p \cap r)$	$p \cap (q \cup r)$
T	T	T	T	T	T	T	T	T	T	T
T	T	F	T	F	F	T	F	F	T	T
T	F	T	F	F	T	T	F	F	T	T
F	T	T	F	T	F	T	F	F	F	F
F	F	T	F	F	F	T	F	F	F	F
F	T	F	F	F	F	T	F	F	F	F
T	F	F	F	F	F	F	F	F	F	F
F	F	F	F	F	F	F	F	F	F	F

Columns 8 and 9 verify the associative law for conjunction. Columns 10 and 11 verify the distributive law for conjunction over disjunction. The reader may extend this table to verify the associative law for disjunction and the distributive law for disjunction over conjunction.

2.7. THE IMPLICATION TRUTH TABLE. The compound statement $\sim(p \cap \sim q)$ is called *implication* and it is designated by $p \to q$, so that

$$\sim(p \cap \sim q) = p \to q.$$

The implication $p \to q$ is to be read "p implies q," or, "if p, then q." Another compound statement that is equivalent to the implication $p \to q$ is $\sim p \cup q$, as is shown in the following implication truth table.

p	q	$\sim p$	$\sim q$	$p \cap \sim q$	$p \to q$ $= \sim(p \cap \sim q)$	$p \to q$ $= \sim p \cup q$
T	T	F	F	F	T	T
T	F	F	T	T	F	F
F	T	T	F	F	T	T
F	F	T	T	F	T	T

The Implication Truth Table

The implication, $p \to q$, is called a *theorem*. If $p \to q$, we say that p is a sufficient condition for q, and that q is a necessary condition for p.

For an example, take the following propositions:

$p = $ "it is raining,"
$q = $ "there is a rain cloud overhead."

Each of the following compound statements would be possible:

a. $p \cap q = $ "it is raining and there is a rain cloud overhead."
b. $\sim p \cap q = $ "it is not raining and there is a rain cloud overhead."
c. $\sim p \cap \sim q = $ "it is not raining and there is not a rain cloud overhead."

However, the following compound statement would not be possible:

    d. $p \cap \sim q$ = "it is raining and there is not a rain cloud over-
head."

The theorem $q \rightarrow p$ is the *converse* of the theorem $p \rightarrow q$.

2.8. THE EQUIVALENCE TRUTH TABLE.   The conjunction
of a theorem, $p \rightarrow q$, with its converse, $q \rightarrow p$, is called the "*equivalence
of p and q*," designated by $p \leftrightarrow q$. Thus,

$$p \leftrightarrow q = (p \rightarrow q) \cap (q \rightarrow p).$$

It follows that

$$p \leftrightarrow q = (\sim p \cup q) \cap (\sim q \cup p)$$
$$= [\sim(p \cap \sim q)] \cap [\sim(q \cap \sim p)].$$

In the equivalence, $p \leftrightarrow q$, we say that "*p* is both a necessary and
a sufficient condition for *q*," or, "*p*, if and only if *q*," or, "iff *p*, then *q*."

The equivalence symbol "$\leftrightarrow$" satisfies the properties of equivalence
relations:

    1.7.1. *Reflexive Property.* $p \leftrightarrow p$.
    1.7.2. *Symmetric Property.* If $p \leftrightarrow q$, then $q \leftrightarrow p$.
    1.7.3. *Transitive Property.* If $p \leftrightarrow q$ and $q \leftrightarrow r$, then $p \leftrightarrow r$.

The theorem $\sim p \rightarrow \sim q$ is the *inverse* of the theorem $p \rightarrow q$. The
theorem $\sim q \rightarrow \sim p$ is the *contrapositive* of $p \rightarrow q$.

				Theorem	Converse	Inverse	Contrapositive	Equivalence
1	2	3	4	5	6	7	8	9
p	q	$\sim p$	$\sim q$	$p \rightarrow q$	$q \rightarrow p$	$\sim p \rightarrow \sim q$	$q \sim q \rightarrow \sim p$	$p \leftrightarrow q$
T	T	F	F	T	T	T	T	T
T	F	F	T	F	T	T	F	F
F	T	T	F	T	F	F	T	F
F	F	T	T	T	T	T	T	T

The Theorem Truth Table

In the previous theorem truth table we include the theorem $p \to q$, its converse, its inverse, its contrapositive, and the equivalence $p \leftrightarrow q$.

A comparison of column 5 with column 8 reveals that

$$p \to q = \sim q \to \sim p,$$

or the theorem is equivalent to its contrapositive. A comparison of column 6 with column 7 reveals that the converse of a theorem is equivalent to the inverse of the theorem.

2.9. TAUTOLOGY.   A compound statement which has the truth value T for every combination of truth values of the propositions which it contains is called a *tautology*. Examples of tautologies are

a. $\sim(\sim p) \leftrightarrow p$.
b. $\sim p \cup (q \to p)$.
c. $\sim q \cup (p \to q)$.

2.10. THE PRINCIPLE OF DUALITY FOR COMPOUND STATEMENTS.   If P, Q are compound statements of propositions involving conjunctions and disjunctions, then the compound statements $\overline{P}, \overline{Q}$ obtained from P and Q, respectively, by replacing every conjunction, $\cap$, by the disjunction, $\cup$, and every disjunction, $\cup$, by the conjunction, $\cap$, are the *dual* compound statements of P and Q.

If the compound statements P and Q are equivalent, then their dual statements $\overline{P}$ and $\overline{Q}$ are equivalent, that is,

$$\text{if } P = Q, \text{ then } \overline{P} = \overline{Q}.$$

It also follows, and can be verified in a truth table, that

$$P \leftrightarrow Q = \overline{P} \leftrightarrow \overline{Q}.$$

For an example, the reader may easily verify in a truth table that, if $P = (p \cup q)$ and $Q = \sim(\sim p \cap \sim q)$, then

$$P = p \cup q = \sim(\sim p \cap \sim q) = Q,$$

and the dual statements

$$\overline{P} = p \cap q = \sim(\sim p \cup \sim q) = \overline{Q}.$$

These equivalent statements and the dual statements are forms of De Morgan's laws, which we shall meet again in Chapter Three.

PROBLEM SET 2.1

1. Construct truth tables for the following:
   a. $p \rightarrow (p \rightarrow q)$.
   b. $(p \cup q) \leftrightarrow (q \cup p)$.
   c. $(p \rightarrow q) \leftrightarrow \sim p \cup q$.
   d. $(p \cap q) \rightarrow [(q \cap \sim q) \rightarrow (r \cap q)]$.
   e. $p \cap q = \sim(\sim p \cup \sim q)$.
   f. $p \cup (q \cap r) = (p \cup q) \cap (p \cup r)$.

2. Prove in a truth table that each of the following is a tautology:
   a. $(p \rightarrow \sim q) \leftrightarrow (q \rightarrow \sim p)$.
   b. $[(p \cap q) \cap \sim q] \rightarrow p$.
   c. $(p \rightarrow q) \leftrightarrow (\sim q \rightarrow \sim p)$.
   d. Make up other examples and prove that they are tautologies.

3. Prove in a truth table that

$$(p \leftrightarrow q) \leftrightarrow (q \leftrightarrow r) = p \leftrightarrow r.$$

Does this statement have any meaning? For example, does it have anything to do with the transitivity of "$\leftrightarrow$?"

4. By using suitable translations for propositions p, q, r, write translations of each of the following theorems into statement language of this chapter.
   a. If ABC is a triangle and ABC is isosceles, then ABC has two equal sides.
   b. The necessary and sufficient condition that triangle ABC be congruent to triangle DEF is that sides of triangle DEF be equal to corresponding sides of triangle ABC.
   c. If a line a is perpendicular to line b, and line c is perpendicular to b, then a is parallel to c.
   d. The necessary and sufficient condition that a positive integer be divisible by 3 is that the sum of its digits be a multiple of 3.

CHAPTER THREE

Universes, Sets, Boolean Algebra

3.1. UNIVERSE.   The collection of the numbers, objects or ideas that are to be used as nouns in a particular discussion is called the *universe* for that discussion. We designate the universe by the symbol *U*. We may also call *U* the *universal set* for that particular discussion. The numbers, objects, or abstract ideas that comprise *U* are called the *elements* of *U*.

The universe for any particular discussion must be clearly defined and understood.

3.2. SET.   If the universal set in a particular discussion produces an offspring collection of its own elements, the offspring collection is called a *set*. We designate sets by capital letters from the beginning of the alphabet. A set *A* in the universe *U* is defined by an *eligibility rule*. The eligibility rule for a set *A* must be clearly defined so that:

   i.   it is possible to examine each element of *U* and decide whether it belongs to *A* or does not belong to *A*;

  ii.   either every element of *U* belongs to set *A* or it does not belong to *A*.

It is obvious that the eligibility rule for a set in a given universe *U* separates the elements of *U* into two sets: the set *A* of those elements of *U* that pass the eligibility rule, and, the set *A'* of elements of *U* that fail to pass the eligibility rule. The set *A'* is the *complement*, or the *complementary set*, of *A* in *U*. Every element in the universal set *U* must belong either to *A* or to its complement *A'*.

The concept of set is one of the most important in the language of mathematics. The word *set* has many synonyms, such as class, collection, aggregation. The concept of set is encountered in every language and every environment. We are accustomed to talk about a set of dishes, a set of sterling silver, a set of false teeth, a stamp collection, a sixpack of pop, a package of frozen peas, a bowl of cherries, a bag of potatoes, a herd of cows, a flock of geese, and many other examples.

The elements in a universe U are designated by lower case letters $a, b, c, \ldots$. We use the verb symbol $\in$, so that "$a \in A$" is read "a is an element of A," or "a belongs to A."

Certain sets have been given definite names. These sets of numbers are in frequent usage either as universal sets or as sets in specified universes. These important sets are

$N =$ the set of all natural numbers, or positive integers;
$I \ =$ the set of all integers, including the positive, negative, and zero;
$R =$ the set of all rational numbers;
$K =$ the set of all real numbers;
$C =$ the set of all complex numbers.

Braces are used either to contain the elements of a set, or to contain the eligibility rule which defines the set. For example, the sets A and B are described by showing all elements that belong to each set:

$$A = \{1, 2, 3, 4\};$$
$$B = \{1, 3, 5, 7, \ldots, 23\}.$$

The sets C and D are defined by stating the universe and the eligibility rule within the braces. The semicolon, or the vertical line segment, is used in either set to mean "such that":

$$C = \{x \; ; 1 < x < 5, x \in K\} = \{x \mid 1 < x < 5, x \in K\};$$
$$D = \{x \; ; x > -2, x \in I\} = \{-1, 0, 1, 2, 3, 4, \ldots\}.$$

In these sets, C is contained in the universal set K of all real numbers. C contains those real numbers that are between 1 and 5. The set D is contained in the universal set I of all integers and D contains all of the integers that are greater than -2.

If the universe has been previously defined and is clearly understood for the sets being defined, then it is not necessary to define the universe for each set.

Other procedures for describing sets will be introduced as it becomes convenient to do so.

3.3. SUBSET.   If the universe U contains sets A and B, and if every element of A belongs to B, then A is a *subset* of B. This is written in mathematics language as "$A \subseteq B$." The entire definition of subset may be written in mathematical symbols as follows:

$$A \subseteq B \leftrightarrow [(x \in A) \rightarrow (x \in B)].$$

The reverse verb symbol, $\supseteq$, means "is an overset of." If A is a subset of B, then B is an overset of A, or

$$B \supseteq A \leftrightarrow A \subseteq B.$$

If both $A \subseteq B$ and $A \supseteq B$, then every element of A is an element of B, and every element of B is an element of A. This can happen only if the sets A and B have precisely the same elements and, consequently, are the same identical set. In this case, A is the same as B, and we write "$A = B$." In mathematics language,

$$A = B \leftrightarrow A \subseteq B \quad \text{and} \quad A \supseteq B.$$

If $A \subseteq B$ and there are elements in B that are not in A, then A is a *proper subset* of B. This is written in mathematics as "$A \subset B$," which is read "A is a proper subset of B."

It follows that, if $A \subseteq B$, then, either $A = B$ or $A \subset B$.

The sets of numbers named in Section 3.2 have the following subset relationship:

$$N \subset I \subset R \subset K \subset C.$$

3.4. NULL SET.   The set that contains no elements is called the *empty set*, or the *null set*. The null set is defined by an eligibility rule for which no element can qualify. The null set is designated by the symbol "$\varnothing$." Examples of eligibility rules for $\varnothing$ include:

$\varnothing = \{x \; ; \; x > 2 \text{ and } x < 1 \text{ and } x \in K\}$;
$\varnothing = \{\text{all the ancestors of Adam, the first man}\}$;

$\emptyset = \{x \; ; \; x$ is irrational and $x \in N\}$;

$\emptyset = \{$the Republican delegates at the Democratic convention$\}$.

There is only one null set, $\emptyset$. It is a subset of every set and $\emptyset$ is a proper subset of every nonempty set. The complement of $\emptyset$ in the universe U is U, and the complement of U is the null set $\emptyset$. In mathematics language this statement is

$$\emptyset' = U \qquad \text{and} \qquad U' = \emptyset.$$

3.5. UNION OF TWO SETS.   Two sets, A and B, in the universe U may be merged into a single set S so that every element in S belongs either to A or to B, or to both A and B. The result of the merger is called the *union* of A with B, and it is designated by "$A \cup B$." Then,

$$S = A \cup B = \{s \; ; \; s \in A \text{ or } s \in B\}.$$

For example, if

$$A = \{1, 2, 3, 5\},$$
$$B = \{1, 2, 4, 5, 6\},$$
$$C = \{2, 4, 6\},$$

then,

$$A \cup B = \{1, 2, 3, 4, 5, 6\},$$
$$A \cup C = A \cup B, \quad \text{and}$$
$$B \cup C = B.$$

The following results follow immediately from definitions:

$$A \cup A = A, \qquad A \cup A' = U,$$
$$A \cup \emptyset = A, \qquad A \cup U = U, \qquad \emptyset \cup \emptyset = \emptyset,$$
$$A \subseteq (A \cup B), \qquad B \subseteq (A \cup B).$$

The following important theorem requires little more than the definitions, so we leave its proof to the reader:

Theorem 3.5.1. $(A \subseteq B) \leftrightarrow (A \cup B = B)$.

3.6. INTERSECTION OF TWO SETS.   The set of all elements which belong to both A and B is called the *intersection* of A and B. The intersection set is designated by "$A \cap B$," so that,

$$A \cap B = \{x ; x \in A \text{ and } x \in B\}.$$

For example, with the same sets A, B, C, defined above in Section 3.5, we have

$$A \cap B = \{1, 2, 5\},$$
$$A \cap C = \{2\}, \quad \text{and}$$
$$B \cap C = \{2, 4, 6\}.$$

The following results follow immediately from definitions:

$$A \cap A = A, \quad A \cap A' = \emptyset,$$
$$A \cap \emptyset = \emptyset, \quad A \cap U = A, \quad \emptyset \cap \emptyset = \emptyset,$$
$$(A \cap B) \subseteq A, \quad (A \cap B) \subseteq B.$$

We leave the proof of the following theorem to the reader:

Theorem 3.6.1. $(A \subseteq B) \leftrightarrow (A \cap B = A)$.

3.7. LAWS FOR UNION AND INTERSECTION OF SETS. The union and the intersection of two sets, A and B, in the universe U satisfy the following laws, as can be easily proved by referring to the definitions:

3.7.1. *Idempotence.* $A \cup A = A; A \cap A = A$.

3.7.2. *Commutativity.* $A \cup B = B \cup A; A \cap B = B \cap A$.

3.7.3. *Associativity.* $A \cup (B \cup C) = (A \cup B) \cup C$;
$A \cap (B \cap C) = (A \cap B) \cap C$.

3.7.4. *Distributive Laws.* $A \cup (B \cap C) = (A \cup B) \cap (A \cup C)$;
$A \cap (B \cup C) = (A \cap B) \cup (A \cap C)$.

3.7.5. *The Principle of Duality for Sets.*   In a statement about sets in a universe U we may form a new statement by replacing every $\cup$ with $\cap$ and every $\cap$ with $\cup$, and at the same time replace U with $\emptyset$ and $\emptyset$ with U, and replace $\subseteq$ with $\supseteq$ and $\supseteq$ with $\subseteq$, as such symbols occur in the statement. Then, the new statement is the *dual statement* of the first. If the first statement is true, then the dual statement is also true. If the statement is false, then its dual is false.

3.8. THE NUMBER OF SUBSETS FOR A SET OF n ELE-MENTS. A set containing a definite number, n, of elements is a *finite set*. It is interesting to count the number of all possible subsets for a given finite set of n elements. This number of subsets is calculated as follows:

	number of subsets
the null set, $\varnothing$, is a subset	$1,$
subsets with one element each	$n,$
subsets with two elements each	$\dfrac{n(n-1)}{2},$
subsets with three elements each	$\dfrac{n(n-1)(n-2)}{3\cdot2\cdot1},$
. . . .	,
subsets with $(n-1)$ elements each	$n,$
subsets with n elements each	$1.$

The total number, T, of all of these subsets is

$$T = 1 + n + \frac{n(n-1)}{2} + \frac{n(n-1)(n-2)}{3\cdot2\cdot1} + \cdots + n + 1.$$

The terms in this sum are the coefficients of the terms in the expansion of the binomial $(a + b)^n$. Consequently, it follows that

$$T = (1 + 1)^n = 2^n.$$

The total number of proper subsets for a finite set of n elements is $(2^n - 1)$.

3.9. NUMBER INTERVALS. Given that the universe is the set, K, of all real numbers, and that $a, b \in K$ and $a \le b$. Then the set of real numbers

$$[a, b] = \{x \,;\, a \le x \le b\}$$

is called the *closed number interval* from a to b. The number a is called the *left end number* and the number b is called the *right end number*. If both end numbers are included in the number interval, it is called a *closed* number interval. The square bracket is used to designate that the end number is included in the number interval. If the end number is not included in the number interval, the parenthesis is used to give this information. Thus, the number interval,

$$(a, b) = \{x \; ; a < x < b\},$$

includes neither end number, so it is called an *open* number interval.

If the left end number belongs to the number interval but the right end number is not included, as in

$$[a, b) = \{x \; ; a \leq x < b\},$$

then it is called a *half closed* number interval. If the left end number is not included, but the right end number is included, as in

$$(a, b] = \{x \; ; a < x \leq b\},$$

then it is called a *half open* number interval.

Since each number interval is a set, we may combine number intervals by union and intersection. For some examples, we have

   i. $[3, 5] \cup [2, 6] = [2, 6]$.
  ii. $[3, 5] \cap (4, 6) = (4, 5]$.
 iii. $(2, 3) \cap (3, 4) = (3, 3) = \varnothing$.
 iv. $(2, 3] \cap [3, 4] = [3, 3] = 3$.
  v. $(-1, 2] \cup (3, 6]$ cannot be expressed as a single number interval.

We may select a certain number interval as a universe. Then, we may have complement number intervals with respect to the universal interval. For example, if

$$U = [-20, 20], \quad A = [2, 5], \quad B = (-3, -1], \quad \text{and}$$

$$C = (-1, 2] \cup (3, 6], \quad \text{then}$$
$$A' = [2, 5]' = [-20, 2) \cup (5, 20],$$
$$B' = (-3, -1]' = [-20, -3] \cup (-1, 20], \quad \text{and}$$
$$C' = [-20, -1] \cup (2, 3] \cup (6, 20].$$

PROBLEM SET 3.1

1.   Given $A = \{a, b, c, d\}$. Make a list of all the subsets of A. How many are there? How many more subsets are there for $B = \{a, b, c, d, e\}$?

2. Given the sets

$$U = \{x \; ; x < 11, x \in N\},$$
$$A = \{2, 4, 6, 8, 10\},$$
$$B = \{1, 3, 5, 7, 9\},$$
$$C = \{2, 3, 7, 9\}.$$

Describe each of the following sets by showing all of its elements:

$$A', \quad B', \quad C', \quad A \cup B, \quad B \cup C, \quad U \cup B, \quad A \cap C,$$
$$A \cup (B \cup C), \quad A \cap (B \cap C), \quad A \cap (B \cup C),$$
$$A \cup (B \cap C), \quad (A \cup B) \cap (A \cup C).$$

3. Assume the universe $U = [-20, +20]$. Compute the resulting number interval for each of the following:

a. $[1, 3] \cup [3, 5] =$;   k. $[2, 4) \cap (3, 5) =$;

b. $[1, 3] \cup [2, 4] =$;   l. $[2, 4] \cap (3, 7) =$;

c. $[1, 7] \cup [2, 5] =$;   m. $[2, 5) \cap (4, 6) =$;

d. $[2, 4) \cup [6, 8] =$;   n. $(2, 4) \cap (4, 8) =$;

e. $(5, 6) \cup [6, 8] =$;   o. $[-8, 8]' =$;

f. $(5, 7) \cup [7, 8] =$;   p. $(4, 12)' =$;

g. $(1, 3) \cup (3, 4) =$;   q. $[-3, 0] \cup [3, 10]' =$;

h. $[1, 3] \cup [2, 4] =$;   r. $[-5, 0] \cap [5, 9]' =$;

i. $[1, 3] \cup [3, 7] =$;   s. $[[-5, 0] \cap [6, 9]]' =$;

j. $[1, 3] \cup [5, 7] =$;   t. $[(-4, 0) \cap (4, 8)]' =$.

3.10. BOOLEAN ALGEBRA OF SETS.   It is assumed that the reader is well acquainted with at least a technical knowledge of the ordinary algebra of real numbers with the operations of ordinary addition and ordinary multiplication. An introduction to various algebras with numbers as elements will be taken up in Chapter Five, but more complete studies of the algebra of real numbers must be left to other courses.

In this section we discuss the boolean algebra of sets, named after the English logician and mathematician George Boole (1815–1864). The reader may be surprised to observe the similarities between this algebra and the algebra of real numbers.

Given the universe U containing the sets $A, B, C, \ldots$. The sets may be combined in pairs by $\cup$, union, and $\cap$, intersection. For every pair, A, B, of sets in U, both $(A \cup B)$ and $(A \cap B)$ are sets in U. These operations satisfy the following laws: 3.7.1. idempotence; 3.7.2. commutativity; 3.7.3. associativity; and 3.7.4. distributive laws.

The null set, $\varnothing$, is the identity set for the operation $\cup$, since, for every A, $A \cup \varnothing = A$. The universe, U, is the identity set for the operation $\cap$, since, for every A, $A \cap U = A$. On the other hand, $A \cap \varnothing = \varnothing$ and $A \cup U = U$.

For certain sets we may have the subset relationship $\subseteq$, or $\subset$. The subset relation $\subseteq$ has the following properties:

3.10.1. *Reflexive.* $A \subseteq A$.

3.10.2. *Antisymmetric.* If $A \subseteq B$ and $B \subseteq A$, then $A = B$; if $A \subseteq B$ but $B \nsubseteq A$, then $A \subset B$.

3.10.3. *Transitive.* If $A \subseteq B$ and $B \subseteq C$, then $A \subseteq C$.

3.10.4. *Consistency.* $A \subseteq B$, $A \cup B = B$, and $A \cap B = A$ are equivalent.

3.10.5. *Universal Bounds.* For every set A in U, $\varnothing \subseteq A \subseteq U$.

For every set A in the universe U there is a complement set A' with the properties that

$$A \cup A' = U, \qquad A \cap A' = \varnothing, \qquad \text{and} \qquad (A')' = A.$$

If $A \cap B = \varnothing$, then A and B have no common elements, and they are *disjoint* sets.

The method of proof in each of the following theorems in the boolean algebra of sets makes use only of definitions and of theorems previously proved. This method of proof is called *formal proof*. The reader will note that a formal proof does not require any recourse to sketches or any other kind of diagram.

Theorem 3.10.6. If $B \subseteq C$, then $(A \cap B) \subseteq (A \cap C)$.

Proof: Since $B \subseteq C$, we know that $B \cup C = C$, so that, by replacing C by $B \cup C$ and by using the distributive law, we have

$$A \cap C = A \cap (B \cup C) = (A \cap B) \cup (A \cap C).$$

Since the set $(A \cap C)$ appears in the first and third members of this set of equalities, it follows that

$$(A \cap B) \subseteq (A \cap C).$$

The converse of Theorem 3.10.6 is not true. If it were true it would be the cancellation law for the operation of intersection.

Theorem 3.10.7.   If $A \subseteq C$ and $B \subseteq C$, then $A \cup B \subseteq C$.

Proof: Since $B \subseteq C$, it follows that $B \cup C = C$; and since $A \subseteq C$, it follows that $A \cup C = C$. Consequently, we have

$$(A \cup B) \cup C = A \cup (B \cup C) = A \cup C = C.$$

From the first and last members of this set of equalities it follows that

$$(A \cup B) \subseteq C.$$

Theorem 3.10.8.   If $B \subseteq C$, then $(A \cup B) \subseteq (A \cup C)$.

Proof: Since $B \subseteq C$ we know that $B \cap C = B$, so that

$$A \cup B = A \cup (B \cap C) = (A \cup B) \cap (A \cup C).$$

From the first and third members of this set of equalities we obtain that

$$(A \cup B) \subseteq (A \cup C).$$

Since the converse of Theorem 3.10.8 is not true there is no cancellation law for the operation of union in the boolean algebra of sets.

Theorem 3.10.9.   The necessary and sufficient conditions for a set B to be the complement of a set A are

$$A \cap B = \emptyset \qquad \text{and} \qquad A \cup B = U.$$

Proof: (1)   That the conditions are necessary follows at once, since, if B is the complement of A, then $B = A'$ and we have that

$$A \cap B = A \cap A' = \emptyset, \quad \text{and}$$
$$A \cup B = A \cup A' = U.$$

(2)   That the conditions are sufficient requires that we prove that if both $A \cap B = \emptyset$ and $A \cup B = U$, then $B = A'$. In the following

chain of equalities we make use of the established properties of A', $\emptyset$, U, and of the distributive laws:

$$\begin{aligned}
B &= B \cap U = B \cap (A \cup A') = (B \cap A) \cup (B \cap A') \\
&= (A \cap B) \cup (B \cap A') = \emptyset \cup (B \cap A') \\
&= (A' \cap A) \cup (A' \cap B) = A' \cap (A \cup B) \\
&= A' \cap U = A'.
\end{aligned}$$

Theorem 3.10.10.   (1) $(A \cup B)' = (A' \cap B')$; and
                      (2) $(A \cap B)' = (A' \cup B')$.

The two parts of Theorem 3.10.10 are called De Morgan's laws for sets, after Augustus De Morgan (1806–1871). The second law is the dual of the first.

Proof: The formal proof of De Morgan's law (1) makes use of Theorem 3.10.9 to show that $(A' \cap B')$ is the complement of $(A \cup B)$. The proof makes repeated use of the distributive laws. In the first place,

$$\begin{aligned}
(A \cup B) \cup (A' \cap B') &= [(A \cup B) \cup A'] \cap [(A \cup B) \cup B'] \\
&= [A' \cup (A \cup B)] \cap [A \cup (B \cup B')] \\
&= [(A' \cup A) \cup B] \cap [A \cup (B \cup B')] \\
&= (U \cup B) \cap (A \cup U) \\
&= U \cap U = U.
\end{aligned}$$

In the second place,

$$\begin{aligned}
(A \cup B) \cap (A' \cap B') &= (A' \cap B') \cap (A \cup B) \\
&= [(A' \cap B') \cap A] \cup [(A' \cap B') \cap B] \\
&= [A \cap (A' \cap B')] \cup [A' \cap (B' \cap B)] \\
&= [(A \cap A') \cap B'] \cup (A' \cap \emptyset) \\
&= (\emptyset \cap B') \cup (A' \cap \emptyset) \\
&= \emptyset \cup \emptyset = \emptyset.
\end{aligned}$$

Since both conditions of Theorem 3.10.9 are satisfied it follows that $(A' \cap B')$ is the complement of $(A \cup B)$, or that

$$(A \cup B)' = (A' \cap B').$$

Since the second of De Morgan's laws is the dual of the first no proof is required to establish that it is true.

PROBLEM SET 3.2

1. State the dual of each of Theorems 3.10.6, 3.10.7, and 3.10.8.

2. Ignore the principle of duality for sets and construct a formal proof for the second of De Morgan's law.

3. Prove the theorem: $(A \subseteq B) \leftrightarrow (B' \subseteq A')$.

4. Construct a formal proof for each of the following. Write the dual statement for each statement.
   a. $(A' \cup B)' = (A \cap B')$;
   b. $A \cup (B \cap C)' = (A \cup B') \cup C'$;
   c. $(A \cup B) \cap (A \cup C) = A \cup (B \cap C)$;
   d. $A \cup (A \cap B) = A$;
   e. $(A \cap B')' = (A' \cup B)$.

5. Construct a formal proof for each of the theorems:
   a. If $A \subseteq C$ and $B \subseteq D$, then $(A \cup B) \subseteq (C \cup D)$.
   b. $(A \subseteq \emptyset) \leftrightarrow (A = \emptyset)$.
   c. $(A \subseteq B) \leftrightarrow (A \cap B' = \emptyset)$.

3.11. A BINARY BOOLEAN ALGEBRA APPLIED TO ELEC-TRICAL CIRCUITS.   There are several boolean algebras, each related in some way to logic or to the algebra of sets. In this section we examine a boolean algebra with only two elements, designated by the symbols 0 and 1. Since there are but two elements it is called a *binary* boolean algebra. This will be applied to electrical circuits consisting of the appropriate wires for conducting electrical current, connectors, and switches. Individual circuits will be designated by lower case letters $a, b, c, \ldots$ A circuit a has the value 1 if electrical current can flow through it, or the value 0 if electrical current cannot flow through it.

A circuit a is represented in a *circuit diagram* by the symbol

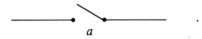

a

The *circuit value* of a is 0 if a is "open" as sketched here. The circuit

value of *a* is 1 if the circuit is closed, which might be represented by
the circuit diagram

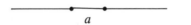

a

However, it is customary to construct circuit diagrams with each circuit
in the "open" situation. Thus, the diagram of circuit *a* will be

a

regardless of whether *a* has the circuit value 0 or the circuit value 1.

Two circuits may be connected together into a new circuit. There
are two different methods for connecting two circuits. The first method
is called *series*. If circuits *a* and *b* are connected in series the result is
designated by $a \cap b$. We may demonstrate the connection of circuits
in circuit diagrams. The circuit diagram for $a \cap b$ is

a   *b*

Series circuit for $a \cap b$

The series circuit $a \cap b$ has the value 1 if and only if both *a* and *b* have
the value 1. The series circuit $a \cap b$ has the value 0 if either of *a* or *b*,
or both *a* and *b*, has the value 0.

The second method of connecting two circuits is called *parallel*. If
circuits *a* and *b* are connected in parallel the result is designated by
$a \cup b$. The circuit diagram for the parallel connection, $a \cup b$, is

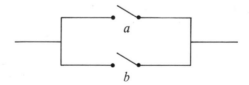

a

b

Parallel circuit for $a \cup b$

The parallel circuit, $a \cup b$, has the value 1 if either *a* or *b*, or both *a*
and *b*, has the value 1. The parallel circuit has the value 0 if and only
if both *a* and *b* have the value 0.

We use the verb symbol "$=$" to mean "has the value of." Thus, "$a = 1$" is read "a has the value of 1," and "$a = b$" is read "a has the value of b."

We denote by a' the circuit which has the value opposite to the value of a. Thus, if a has the value 1, then a' has the value 0, and if a has the value 0, then a' has the value 1. It follows that for every circuit a, the identities

$$a \cap a' = 0 \qquad \text{and} \qquad a \cup a' = 1$$

are true. For every circuit a and any circuit b the following identities are true:

$$(a \cap a') \cap b = 0, \qquad (a \cup a') \cap b = b,$$
$$(a \cap a') \cup b = b, \qquad (a \cup a') \cup b = 1.$$

The information about circuit values for series and parallel circuits may be summarized in the following "multiplication" tables:

	b					b	
$\cap$	0	1			$\cup$	0	1
0	0	0			0	0	1
1	0	1			1	1	1

       $a \cap b$ $a \cup b$

(with a labeling the left rows of each table)

When two, or more, circuits are combined we may compute the values in a circuit value table; for example,

a	b	a'	b'	$a \cap b$	$a \cup b$	$a' \cap b'$	$a' \cup b'$	$(a \cap b)'$
1	1	0	0	1	1	0	0	0
1	0	0	1	0	1	0	1	1
0	1	1	0	0	1	0	1	1
0	0	1	1	0	0	1	1	1

A Circuit Value Table

The laws for combining circuits are similar to those for the algebra of sets:

3.11.1. *Idempotence.* $a \cap a = a;\ a \cup a = a$.

3.11.2. *Commutativity.* $a \cap b = b \cap a;\ a \cup b = b \cup a$.

3.11.3. *Associativity.* $(a \cap b) \cap c = a \cap (b \cap c)$;
$(a \cup b) \cup c = a \cup (b \cup c)$.

3.11.4. *Distributive Laws.* $a \cap (b \cup c) = (a \cap b) \cup (a \cap c)$;

$a \cup (b \cap c) = (a \cup b) \cap (a \cup c)$.

We leave to the reader the task of proving these laws in a circuit value table.

3.11.5. *The Principle of Duality for Electrical Circuits.* In any statement about series and parallel connections of electrical circuits in which the verb $=$, meaning "has the value of," is used the series symbol $\cap$ may be interchanged with the parallel symbol $\cup$, and, simultaneously, if they occur, the values 0 and 1 are to be interchanged. The resulting statement is the dual of the first, and if the first is true then its dual statement is true.

3.11.6. *De Morgan's Laws.* $(a \cap b)' = a' \cup b'$;
$(a \cup b)' = a' \cap b'$.

Both of De Morgan's laws are easily proved in a circuit value table, or the first may be proved and the truth of the second will follow from the principle of duality.

A circuit equation may be illustrated in a circuit diagram. For examples, the circuit diagrams for the two distributive laws are sketched here.

Circuit diagram for $a \cap (b \cup c) = (a \cap b) \cup (a \cap c)$

Circuit diagram for $a \cup (b \cap c) = (a \cup b) \cap (a \cup c)$

Complicated connections of several circuits may be simplified by means of the binary Boolean algebra for electrical circuits. For an example, the circuit

$$[(a \cup b) \cap (a' \cup c)] \cup [b \cap (b' \cup c)]$$

may be simplified by algebraic manipulations as follows:

$[(a \cup b) \cap (a' \cup c)] \cup [b \cap (b' \cup c)]$
$= [(a \cup b) \cap a'] \cup [(a \cup b) \cap c] \cup [(b \cap b') \cup (b \cap c)]$
$= [(a' \cap a) \cup (a' \cap b)] \cup [(a \cap c) \cup (b \cap c)] \cup (b \cap c)$
$= (a' \cap b) \cup (a \cap c) \cup (b \cap c)$
$= (a' \cap b) \cup [(a \cup b) \cap c].$

The circuit diagram showing the first and final members of this chain of equalities is given here.

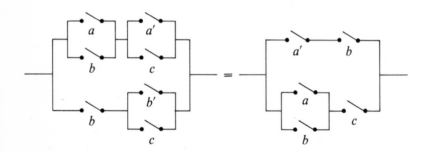

PROBLEM SET 3.3

1. Prove both of De Morgan's laws for circuits in a circuit value table.

2. In each of the following combination circuits, reduce, by algebraic manipulation, the first member to the second member. Show the equivalence in a circuit value table. Construct the circuit diagram for each member.

    a. $(a \cup b) \cup (a \cap c) = a \cup b$;
    b. $a \cap (a' \cup b) = a \cap b$;
    c. $a \cup (a' \cap b) = a \cup b$;
    d. $(a \cup b) \cap (a' \cup c) \cap (b \cup c) = (a \cup b) \cap (a' \cup c)$;

e. $[(a \cap b) \cup c']' \cup [(a' \cap c) \cup b] = [(a' \cup c) \cap c] \cup b$
$$= [(a' \cup b') \cap c] \cup b.$$

3.   Express the algebraic form indicated by the following circuit diagram. Simplify the algebraic expression. Construct the new diagram which is equivalent to the one given.

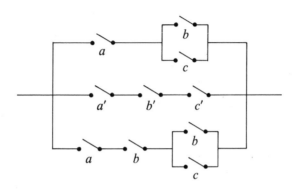

3.12. EQUIVALENT SETS.   Given two sets, A and B, each with distinct elements. If it is possible to correlate the elements of A with the elements of B such that for every $a \in A$ there is one and only one image element $b \in B$, and every element $b \in B$ is the image for one and only one element $a \in A$, then we can set up a 1 : 1 correspondence, read "one-to-one correspondence," between the sets A and B. In general, if a 1 : 1 correspondence can be set up between two sets, A and B, then it can be done in many ways.

If we can establish a 1 : 1 correspondence between two sets A and B, then we say that the two sets are *equivalent*. We shall use the tilde, $\sim$, as the verb symbol to mean "is equivalent to," or "has the same power as." Thus, $A \sim B$ is read either "A is equivalent to B" or "A has the same power as B." Another translation is "A has the same potency as B." Regardless of the English words used for the verb symbol "$\sim$," when the nouns are sets, the meaning is the same: that it is possible to establish a 1 : 1 correspondence between the elements of A and the elements of B.

It is easy to see that this new verb symbol, $\sim$, satisfies the properties of an equivalence verb:

3.12.1. *Reflexive Property.* $\forall\, A,\, A \sim A.$

3.12.2. *Symmetric Property.*   If $A \sim B$, then $B \sim A$.

3.12.3. *Transitive Property.*   If $A \sim B$ and $B \sim C$, then $A \sim C$.

3.13. CARDINAL NUMBER.   It is obvious that A may be equivalent to B even though the elements of A may be very different from the elements of B. (For example, the elements of A may be chairs and the elements of B may be persons sitting in the chairs.) For a given set A, those sets which are equivalent to A may vary widely in the appearence, or character, of their elements. However, since any two sets in this set of sets are equivalent, there must be a common characteristic shared by these sets. We call this common characteristic shared by two equivalent sets the *cardinal number*, or the power, or the potency, of the sets. We shall use the term "cardinal number," so that, if $A \sim B$, the sets A and B have the same cardinal number. Consequently, $A \sim B$ may be read, also, as "A has the same cardinal number as B."

If A and B are finite sets, then the cardinal number is the number of elements, since any two finite sets having the same number of elements can be put into 1 : 1 correspondence. For example, if A is a finite set containing 10 elements, then any other finite set containing 10 elements is equivalent to A, and all of these sets have the same cardinal number, 10. The real importance of the concept of cardinal number becomes more significant when we discuss the cardinal numbers for sets that are not finite.

3.13'. INFINITE SET.   If, for a given set A, it is possible to find a proper subset, B, which is also equivalent to A, that is, if it is possible to have both $B \subset A$ and $B \sim A$, then A is an *infinite set*. A set that is not an infinite set is a *finite set*.

For an example of an infinite set, consider the set, N, of natural numbers and the set, E, of all even natural numbers,

$$E = \{2, 4, 6, \ldots\}.$$

Since every element of E is a natural number, and since there are natural numbers in N which are not in E (the odd natural numbers), it follows that E is a proper subset of N. We may establish a 1 : 1 correspondence between the elements of E and the elements of N simply by letting the image of each $n \in N$ be $2n \in E$, as is demonstrated in the following scheme:

$$N = \{1, \quad 2, \quad 3, \quad . \quad . \quad ., \quad n, \quad . \quad . \quad .\} \quad ,$$
$$E = \{2, \quad 4, \quad 6, \quad . \quad . \quad ., \quad 2n, \quad . \quad . \quad .\} \quad .$$

Since we have established that both $E \subset N$ and $E \sim N$, it follows that N is an infinite set. Moreover, since $E \sim N$, it follows that E is also an infinite set.

The cardinal number of the set N, and of all infinite sets equivalent to N, has been arbitrarily assigned the name $\aleph_0$, read "aleph-null." In the next three sections of this chapter we shall discuss examples of infinite sets with the cardinal number $\aleph_0$.

The set K of all real numbers is an infinite set which can be proved not to be equivalent to N. The set K is said to have greater potency than the set N. The cardinal number of K, and of all sets equivalent to K, has been assigned the name "c." There are other infinite sets which are equivalent neither to N nor to K. The cardinal numbers of infinite sets are elements in an infinite set, which is called the set of *transfinite* numbers.

The reader is warned that he should be extremely careful in mathematics in the use of the word "infinite." The definition of infinite set just given is the only acceptable definition in this course. Such meaningless expressions as "infinite number," "infinitely many," "infinitely few," and so on will be regarded as utter nonsense and will not be tolerated in mathematics language.

If a given set A is not an infinite set, then it is a finite set. The cardinal number of the null set is 0. The cardinal number of a finite set containing n elements is $n \in N$.

3.14. COUNTING, COUNTABLE SET. We assume that N is the symbol, not only for the natural numbers, but also for the natural numbers in the natural order. It is assumed that the reader needs no further description of what is meant by the natural order of the natural numbers.

For a certain $q \in N$ we define the set N_q to consist of the first q natural numbers, that is,

$$N_q = \{1, 2, 3, \ldots, q\}.$$

The cardinal number of N_q is q.

The process of establishing a 1 : 1 correspondence between the elements of a set A and the elements of N is called *counting*. If A is a finite set, then there will be a number, q, such that A is equivalent to N_q. If A is an infinite set and its elements can be put into 1 : 1 correspondence with the natural numbers N, then A is called a *countable* infinite set, or a *denumerable* infinite set, or an *enumerable* infinite set.

Every finite set is countable and its cardinal number is the number of elements that it contains. Not every infinite set is countable. The cardinal number of each infinite countable set is $\aleph_0$, the cardinal number of N.

An infinite set that cannot be counted is called a *noncountable* set, or a *nondenumerable* set, or a *nonenumerable* set. Every noncountable set is an infinite set.

3.15. COUNTABILITY OF THE SET OF POSITIVE RATIONAL NUMBERS.

A number which can be put into the form p/q, where $p, q \in N$, is called a *positive rational number*. Whenever a rational number is represented in decimal form, it is a repeating decimal. For example, $1/3 = 0.333\ldots$, $1/2 = 0.5000\ldots = 0.4999\ldots$, and so on. The set of all positive rational numbers is designated by the symbol R_+, where

$$R_+ = \{p/q \; ; p, q \in N\}.$$

We propose to demonstrate that the set R_+ is countable. The first method of counting R_+ is based on the method proposed by Georg Cantor (1845–1918). The elements of R_+ are arrayed in horizontal rows and vertical columns. In the first row are placed the positive rational numbers with numerator 1 and in the natural order of the denominators. In the second row are placed the positive rational numbers with numerator 2 and arranged in the natural order of the denominators. This is continued. If $p \in N$, the positive rational numbers with numerator p are placed in the pth row of the array in the natural order of the denominators. When we look at the columns in the array we find that the first column contains all of the positive rational numbers with denominator 1 and arranged in the natural order of the numerators, the second column contains those with denominator 2, and so on. The qth column will contain the positive rational numbers with denominator q and arranged in the natural order of the numerators. The rational number p/q will be found in the pth row and qth column. Every positive rational number occurs at least one time in the Cantor array. If we amend the definition of positive rational number to be more in line with common usage we define a positive rational number p/q to be the set of numbers $\{pk/qk \; ; p, q, k \in N, p$ and q have no common divisor$\}$. With this definition, every positive rational number occurs many times in the Cantor array. If we can show that all of the rational numbers included in the Cantor array can be counted, we certainly shall have demonstrated the "smaller" set, R_+, is countable.

The Cantor array of the positive rational numbers

If we follow the arrows sketched in the Cantor array we shall have set up a 1 : 1 correspondence between the elements of R_+ and the elements of N, which will start as follows:

$$R_+ = \{1/1, 1/2, 2/1, 3/1, 2/2, 1/3, 1/4, 2/3, \ldots\},$$
$$N = \{\ 1\ ,\ 2\ ,\ 3\ ,\ 4\ ,\ 5\ ,\ 6\ ,\ 7\ ,\ 8\ ,\ldots\}.$$

If this process of counting along the diagonals is continued it will establish that for each rational number in the array there is a corresponding natural number, and for every natural number there will be one rational number in the array. Consequently, the set R_+ is equivalent to N, and we may conclude that R_+ is a countable infinite set with cardinal number $\aleph_0$.

A second method for enumerating the rational numbers in R_+ is somewhat similar to the Cantor array, but allows us to count only the positive rational numbers such that no two of them are equal. We define the *height* of the positive rational number, p/q, to be $(p + q)$ if p and q have no common integral divisor, or, if p and q have the common divisor $k \in N$, so that $p/q = kr/ks$, where r and s have no common divisor, then the height of p/q is $(r + s)$. If we designate the height by h, then,

$$h = p + q; \qquad \text{or, if } p + q = kr + ks, \text{ then } h = r + s.$$

For each height, h, there can be only a finite set of rational numbers. Since every finite set is countable, it must follow that after we have counted a finite set, N_q, we may "add on" another finite set of p elements by continuing to count. The result will be the counting of the finite set $N_q + N_p = N_{q+p}$. In the following table, the first column contains the successive values of h; the second column contains the set of positive rational numbers $p/q = kr/ks$, where $p, q, k \in N$, which can have each height h; the third column contains the positive rational numbers actually contained in the second column; the fourth column contains the counting numbers for the positive rational numbers in column three. If this table is continued, it is obvious that we shall find that each height, h, adds a finite set of positive rational numbers to the list of those already counted. Consequently, regardless of how long this table is extended, the positive rational numbers appearing in the third column will be counted. On the other hand, every positive rational number must be included in the third column corresponding to its height in the first column. It follows that, again, we have established a 1 : 1 correspondence between the elements of R_+ and the elements of N.

h	p/q	ELEMENTS OF R_+ WITH HEIGHT h	COUNTING NUMBERS
1	None	None	
2	k/k	1	1
3	$2k/k$	2	2
	$k/2k$	1/2	3
4	$3k/k$	3	4
	$k/3k$	1/3	5
5	$4k/k$	4	6
	$3k/2k$	3/2	7
	$2k/3k$	2/3	8
	$k/4k$	1/4	9
6	$5k/k$	5	10
	$k/5k$	1/5	11
7	$6k/k$	6	12
	$5k/2k$	5/2	13
	$4k/3k$	4/3	14
	$3k/4k$	3/4	15
	$2k/5k$	2/5	16
	$k/6k$	1/5	17

For another version of counting the positive rational numbers the reader should study an interesting article published in 1948 (in *The American Mathematical Monthly* **55**, No. 2, pages 65–70) entitled "Denumerability of the Rational Number System," by Professor Leon S. Johnston, who taught at the University of Detroit from 1929 until his death in 1957.

3.16. COUNTABILITY OF THE SET OF ALGEBRAIC NUMBERS. In ordinary algebra, an expression with the form

$$P(x) = a_0x^n + a_1x^{n-1} + a_2x^{n-2} + \cdots + a_{n-1}x + a_n,$$

where $n \in N$ and each of the coefficients $a_0, a_1, a_2, \ldots, a_n$ is an integer $\in I$, is called a *polynomial* of degree n in the variable x. The open statement

$$P(x) = 0$$

is called an *algebraic equation*.

Any number, r, for which $P(r) = 0$ is called a *zero* of the polynomial $P(x)$, or r is called a *root* of the algebraic equation $P(x) = 0$.

A number, r, that can be one of the zeros of some polynomial $P(x)$, or one of the roots of an algebraic equation $P(x) = 0$, is called an *algebraic number*.

The set, A, of all algebraic numbers contains as proper subsets:

a. the set N of natural numbers, since every $n \in N$ is a zero of the polynomial $(x - n)$;

b. the set I of all integers, since every $a \in I$ is a zero of the polynomial $(x - a)$;

c. the set R of all rational numbers, p/q, where $p, q \in I$, $q \neq 0$, since every $p/q \in R$ is a zero of the polynomial $(qx - p)$;

d. many of the irrational real numbers, such as $\sqrt{2}$, which is a zero of the polynomial $(x^2 - 2)$, and, $3 + \sqrt{5}$, which is a zero of the polynomial $(x^2 - 6x + 4)$, the number $(\sqrt[3]{3} - 7)$, which is a zero of the polynomial $(x^3 + 21x^2 + 147x + 343)$, and many, many other irrational numbers, particularly those irrational numbers that are combinations of radicals.

e. complex numbers of the form $a + bi$, where $i^2 = -1$, and $a, b \in R$, and many other complex numbers.

Before we proceed to show that the set, A, of all algebraic numbers is countable, we must define the *absolute value* of a real number $k \in K$, designated by the symbol $|k|$. This important concept is defined as follows:

$$|k| = k, \qquad \text{if} \quad k \geq 0,$$
$$= -k, \qquad \text{if} \quad k < 0.$$

Thus, for examples, $|3| = 3$, $|-7| = 7$, $|-3.14| = 3.14$.

We define the *height*, h, of the polynomial $P(x)$, defined at the beginning of this section, to be

$$h = n + |a_0| + |a_1| + |a_2| + |a_3| + \cdots + |a_{n-1}| + |a_n|.$$

Every polynomial, $P(x)$, has a height $h \in N$. For every natural number, except 1, there exists at least one polynomial having that natural number for height. For $h = 1$ there is no polynomial. For $h = 2$ there are two polynomials, x and $-x$. The polynomials with $h = 3$ are x^2, $-x^2$, $x - 1$, $x + 1$, $-x - 1$, $-x + 1$, $2x$, $-2x$. No matter what $h \in N$ there will be a finite set of polynomials with height h. In the following table we list the first four possible heights in the first column. In the second column we list the polynomials with height h, and in the third column we list the algebraic numbers that are zeros of the polynomials in the second column, with the exception that any algebraic number that has already been listed above will not be listed again.

Since, for each height h, there can be only a finite set of polynomials, there will be only a finite set of new algebraic numbers added for each height. The counting numbers listed in the third column may be extended to the newly added algebraic numbers for every successive height h.

h	POLYNOMIALS WITH HEIGHT h	ALGEBRAIC NUMBERS	COUNTING NUMBERS
1	None	None	
2	$x, -x$	0	1,
3	$x^2, -x^2, x - 1, x + 1,$ $-x + 1, -x - 1,$ $2x, -2x$	$1, -1$	2, 3,
4	$x - 2, x + 2, 2x - 1,$ $2x + 1, 3x, -3x,$ $x^2 - 1, x^2 + 1,$ $x^2 - x, x^2 + x$	$2, -2, 1/2,$ $-1/2, i,$ $-i$	4, 5, 6, 7, 8, 9,

Algebraic Numbers for Polynomials with $h = 1, 2, 3, 4$

If this table is continued, it will establish a 1 : 1 correspondence between the elements of N and the elements of the set A of algebraic numbers. Since every polynomial will be included in column two, then every algebraic number will be found in the third column. No matter how far we have counted the algebraic numbers as indicated in the fourth column, we can always proceed with the counting process to include the next finite set. It follows, then, that the set A of all algebraic numbers is countable. Since $A \sim N$, then A is a countable infinite set.

3.17. SOME STATEMENTS ABOUT COUNTABLE INFINITE

SETS.   We summarize the present study of countable infinite sets by making the following statements. These statements are proved in more complete treatments of the theory of sets.

3.17.1.   A countable set of countable sets is countable.

3.17.2.   A countable infinite set contains a countable set of countable infinite subsets.

3.17.3.   The union of a countable set of countable sets is countable.

3.17.4.   The set of all positive prime numbers is a countable infinite set.

3.17.5.   The set of all triangular numbers is a countable infinite set.

3.17.6.   The set of all square numbers is a countable infinite set.

3.17.7.   The set of all perfect numbers is a countable infinite set.

The cardinal number of any countable set is $\aleph_0$.

3.18. NONCOUNTABLE SETS.

An infinite set which cannot be counted is called a *noncountable infinite set*, or, since every finite set is countable, we may say more briefly, *noncountable set*.

Since the proof of noncountability is considerably more subtle than the reader needs to experience at this time, such a proof will not be included here. The set of all real numbers is noncountable and the cardinal number of K is assigned the symbol c. The set K has greater potency than the set N.

A real number that is not an algebraic number is called a *transcendental number*. If we designate the set of real algebraic numbers by A_k and the set of real transcendental numbers by T, then we have

$$A_k \cup T = K \qquad \text{and} \qquad A_k \cap T = \varnothing.$$

Since $A_k \subset A$, it follows that A_k is countable. Consequently, it must follow that the set T of transcendental numbers is noncountable.

If we designate the set of all irrational numbers by M, and the set of all rational numbers by R, then

$$R \cup M = K \quad \text{and} \quad R \cap M = \varnothing.$$

Since R is countable, it follows that M is noncountable. Each of T and M has the cardinal number c.

PROBLEM SET 3.4

1. Prove that the set I of all integers is countable.

2. Prove that the set of all rational numbers, including the positive rational numbers, the negative rational numbers, and 0, is countable.

3. Find the polynomials, and as many as you can conveniently, of the algebraic numbers arising from polynomials for which $h = 5$ and $h = 6$.

4. Given that $a, b, c, d \in N$ and that $a/b < c/d$ is defined to mean that $ad < bc$. Prove that there is a rational number r/s, where $r, s \in N$, which is between a/b and c/d, so that $a/b < r/s < c/d$. Find a general formula for determining every rational number that lies between a/b and c/d. By means of this formula prove that between any two positive rational numbers $a/b < c/d$ there is an infinite countable set of positive rational numbers.

3.19. PROPERTIES OF THE SET N OF NATURAL NUMBERS. Since the set N of natural numbers is used so frequently and so commonly by everyone, it is usually taken for granted that there is no need for describing its properties. In fact, Leopold Kronecker (1823–1891), a German mathematician, was quoted as having said, "God made the integers, all the rest is the work of man."

The Italian mathematician Giuseppi Peano (1858–1932) proposed the following five axioms to describe the properties of the set N of natural numbers:

3.19.1. *Axiom 1.* 1 is a natural number.

That is, N is not an empty set, since it contains an element 1, called "one."

3.19.2. *Axiom 2.* For each $x \in N$ there exists exactly one other element $x' \in N$, which is called the *successor* of x.

Thus, if $x = y$ and $x, y \in N$, then $x' = y'$ and $x', y' \in N$.

3.19.3. *Axiom 3.* We always have $x' \neq 1$.

That is, there is no natural number whose successor is 1. In this respect, 1 is the *first* natural number. In fact, it is not difficult to prove that in any subset of N there is always a first natural number.

3.19.4. *Axiom 4.* If $x' = y'$, then $x = y$.

That is, for any given number there exists either no number or exactly one number whose successor is the given number.

3.19.5. *Axiom 5* (The Axiom of Induction). Given a set of natural numbers $M \subseteq N$ with the following properties:

    a.   1 belongs to M;
    b.   if $x \in M$, then $x' \in M$.

Then M contains all the natural numbers, $M = N$.

In our usual notation, of course, x', the successor to x, is written $x + 1$.

3.20. THE METHOD OF MATHEMATICAL INDUCTION. Axiom 5 of the Peano axioms for the set N of natural numbers has an important application to proving the truth of certain mathematical statements. The procedure of proof based on this axiom is called the method of *mathematical induction*, or the method of complete induction. This method of proof is based on the following version of Axiom 5:

A statement is true for every natural number $n \in N$ if the following conditions are satisfied:

Condition a.   The statement is true for $n = 1$.
Condition b.   The truth of the statement for any natural number $n = k$ implies its truth for the succeeding natural number, $n = k + 1$.

For examples, we use the method of mathematical induction to prove each of the following theorems:

3.20.1. *Theorem* 1.   The sum, S_n, of the first n natural numbers is

$$S_n = \frac{n(n+1)}{2}.$$

Proof: Let M be the set of natural numbers for which the theorem is true, so that, for every $m \in M$,

$$S_m = 1 + 2 + 3 + \cdots + m = \frac{m(m+1)}{2}.$$

Then, condition a is fulfilled, since $S_1 = 1 = 1(1+1)/2 = 1$, so $1 \in M$; and condition b is fulfilled, since if $k \in M$, and

$$S_k = 1 + 2 + 3 + \cdots + k = \frac{k(k+1)}{2},$$

we derive that

$$S_{k+1} = 1 + 2 + 3 + \cdots + k + (k+1) = S_k + (k+1)$$
$$= \frac{k(k+1)}{2} + (k+1) = \frac{(k+1)(k+2)}{2},$$

so $(k+1) \in M$. Consequently, $M = N$.

3.20.2. *Theorem* 2.   The sum of the first n odd natural numbers is
$$S_n = n^2.$$

Proof.   Let M be the set of natural numbers for which the theorem is true, so that for every $m \in M$, we have that

$$S_m = 1 + 3 + 5 + \cdots + (2m-1) = m^2.$$

Since $S_1 = 1 = 1^2$, it follows that $1 \in M$ and condition a is fulfilled. If $k \in M$, then it is true that

$$S_k = 1 + 3 + 5 + \cdots + (2k-1) = k^2.$$

From this we derive that

$$S_{k+1} = S_k + (2k+1) = k^2 + (2k+1) = (k+1)^2,$$

and that $(k+1) \in M$. It follows that condition b is fulfilled. Consequently, $M = N$.

3.20.3. *Theorem* 3.   Prove that

$$S_n = \frac{1}{1\cdot2} + \frac{1}{2\cdot3} + \frac{1}{3\cdot4} + \cdots + \frac{1}{n(n+1)} = \frac{n}{n+1}.$$

Before we start the proof we shall pause to see how the value of S_n may be derived. We observe that each of the terms of S_n may be written as the difference between two fractions:

$$S_n = \left(\frac{1}{1} - \frac{1}{2}\right) + \left(\frac{1}{2} - \frac{1}{3}\right) + \left(\frac{1}{3} - \frac{1}{4}\right) + \cdots$$

$$+ \left(\frac{1}{n-1} - \frac{1}{n}\right) + \left(\frac{1}{n} - \frac{1}{n+1}\right)$$

$$= \frac{1}{1} - \frac{1}{n+1} = \frac{n}{n+1}.$$

Proof. Let M be the set of natural numbers for which the theorem is true, so that for every $m \in M$, $S_m = m/(m+1)$. Since

$$S_1 = \frac{1}{1\cdot2} = \frac{1}{1+1} = \frac{1}{2},$$

it follows that $1 \in M$. For $k \in M$,

$$S_k = \frac{k}{k+1}, \quad \text{and}$$

$$S_{k+1} = S_k + \frac{1}{(k+1)(k+2)} = \frac{k}{(k+1)} + \frac{1}{(k+1)(k+2)}$$

$$= \frac{k^2 + 2k + 1}{(k+1)(k+2)} = \frac{(k+1)^2}{(k+1)(k+2)} = \frac{k+1}{k+2},$$

and it follows that $(k+1) \in M$. Consequently, $M = N$.

3.20.4. *Theorem* 4.   Prove that for every $n \in N$,

$$(a+b)^n = a^n + na^{n-1}b + \frac{n(n-1)}{2}a^{n-2}b^2 + \frac{n(n-1)(n-2)}{3\cdot2}a^{n-3}b^3$$

$$+ \cdots + \frac{n(n-1)}{2}a^2b^{n-2} + nab^{n-1} + b^n.$$

Proof: Let M be the set of natural numbers for which the theorem is true.

Since $(a + b)^1 = a + b$, which is in the pattern of the formula to be proved, it follows that $1 \in M$, and that condition a is fulfilled.

Next, if $k \in M$, then

$$(a + b)^k = a^k + ka^{k-1}b + \frac{k(k-1)}{2} a^{k-2}b^2 + \frac{k(k-1)(k-2)}{3\cdot2} a^{k-3}b^3$$

$$+ \cdots + \frac{k(k-1)}{2} a^2b^{k-2} + kab^{k-1} + b^k.$$

From this we derive by actual multiplication that

$$(a + b)^{k+1} = (a + b)(a + b)^k = a^{k+1} + (k+1)a^kb + \frac{(k+1)k}{2} a^{k-1}b^2$$

$$+ \frac{(k+1)k(k-1)}{3\cdot2} a^{k-2}b^3 + \cdots + (k+1)ab^k + b^{k+1}.$$

It follows that if $k \in M$, then $k + 1 \in M$. Consequently, $M = N$.

PROBLEM SET 3.5

Prove each of the following statements by the method of mathematical induction:

1. The sum of the squares of the first n natural numbers is
$$\frac{n(n + 1)(2n + 1)}{6}.$$

2. The sum of the squares of the first n odd natural numbers is
$$\frac{n(2n - 1)(2n + 1)}{3}.$$

3. The sum of the cubes of the first n natural numbers is
$$\frac{n^2(n + 1)^2}{4}.$$

4. $1\cdot2 + 2\cdot3 + 3\cdot4 + \cdots + n(n + 1) = \frac{n(n + 1)(n + 2)}{3}.$

5. $\frac{1}{1\cdot4} + \frac{1}{4\cdot7} + \frac{1}{7\cdot10} + \cdots + \frac{1}{(3n - 2)(3n + 1)} = \frac{n}{3n + 1}.$

CHAPTER FOUR

Relations, Functions, and Graphs

4.1. THE CARTESIAN LINE. We assume that the reader has some intuitive knowledge about the character of the set K of real numbers. The development of the properties of real numbers and the structure of the set K of all real numbers must be left for other courses.

One of the useful facts in this area of study is that the set K of all real numbers is equivalent to the set of points on a line. We set up the $1 : 1$ correspondence between the numbers in K and the points on the line by first selecting a point on the line to correspond to the number 0, and then selecting a convenient distance unit, which is measured successively in both directions from the point corresponding to 0. If we place the line in a "horizontal" position, then the first point one unit to the right of 0 will correspond to the number 1, the second point to the right will correspond to the number 2, and so on for all of the natural numbers. The first point one unit distance to the left of 0 will correspond to the number -1, the second point to the left will correspond to -2, and so on for all of the negative integers. The points on the line between 0 and 1 will correspond to the real numbers between 0 and 1, the points on the line between 1 and 2 will correspond to the real numbers between 1 and 2, and so on. Thus, a $1 : 1$ correspondence between the real numbers and the points on the line will have been established. We say that we have *mapped* the real numbers onto the line.

The line on which the real numbers have been mapped as just described is called a *cartesian line*. The name cartesian gives credit to the French mathematician René Descartes (1596–1650).

··· −5 −4 −3 −2 −1 0 1 2 3 4 ···

The Cartesian Line

4.2. THE CARTESIAN PRODUCT OF TWO SETS. Given two sets of distinct elements, A and B. The set of all ordered pairs, (a, b) in which $a \in A$ and $b \in B$, is called the *cartesian product*, $A \times B$. Symbolically, the cartesian product $A \times B$ is defined as follows:

$$A \times B = \{(a, b) \; ; a \in A \text{ and } b \in B\}.$$

For an example, if $A = \{1, 2, 3, 4\}$ and $B = \{p, q, r\}$, then

$$A \times B = \{(1, p), (1, q), (1, r), (2, p), (2, q), (2, r),$$
$$(3, p), (3, q), (3, r), (4, p), (4, q), (4, r)\}.$$

On the other hand,

$$B \times A = \{(p, 1), (p, 2), (p, 3), (p, 4),$$
$$(q, 1), (q, 2), (q, 3), (q, 4),$$
$$(r, 1), (r, 2), (r, 3), (r, 4)\}.$$

It is quite obvious that the cartesian product is not commutative,

$$B \times A \neq A \times B.$$

The two factors in a cartesian product need not be different sets. In fact, it is quite common to have the cartesian product of a set with itself. The most commonly used cartesian product is that in which both sets are real numbers, as will be discussed in Section 4.3.

The cartesian product of any set with the null set is the null set,

$$A \times \varnothing = \varnothing \times A = \varnothing.$$

Conversely, if $A \times B = \varnothing$, then either $A = \varnothing$, or $B = \varnothing$, or $A = B = \varnothing$.

The cartesian product of three sets A, B, C is interpreted as follows:

$$A \times B \times C = A \times (B \times C) = (A \times B) \times C$$
$$= \{(a, b, c) \; ; a \in A, b \in B, c \in C\}.$$

Thus, $A \times B \times C$ results in the set of *ordered triples* of elements in which the first elements are from set A, the second elements belong to set B, and the third elements are members of set C. In Chapter Eight we shall discuss some of the applications of the set of ordered triples of real numbers.

PROBLEM SET 4.1

1. Given the sets $A = \{1, 2, 3, 4\}$, $B = \{p, q, r\}$, $C = \{r, s, t\}$. Calculate the result of each of the following:

    a. $A \times (B \cup C)$;

    b. $A \times (B \cap C)$;

    c. $(A \times B) \times C$;

    d. $[(A \cup B) \cup C] \times C$.

2. Find an example in which $A \times B = B \times A$. Determine conditions under which $A \times B = B \times A$.

3. Given sets A, B, C, D. Prove formally, or in word arguments:

    a. $(A \cup B) \times C = (A \times C) \cup (B \times C)$;

    b. $(A \cap B) \times (C \cap D) = (A \times C) \cap (B \times D)$.

4.3. RELATION.   The cartesian product of the set of real numbers with itself is designated by the symbol K_2, where

$$K_2 = K \times K = \{(x, y) \; ; \; x, y \in K\}.$$

The set K_2 is usually called "the set of ordered pairs of real numbers."

A subset of K_2 is called a *relation*. Thus, any set of ordered pairs in which the elements are real numbers is a relation. A relation may consist of a finite set of ordered pairs of real numbers, or an infinite set of ordered pairs of real numbers.

4.4. THE CARTESIAN PLANE.   Given two cartesian lines with the same distance unit, we place the cartesian lines on a plane so that they intersect at right angles at the point which corresponds to the number 0 on each line. We select one of the cartesian lines to be the *horizontal axis* and the other one to be the *vertical axis*. We use the variable x to represent the real numbers on the horizontal axis, which will now be known as the "x axis," or the axis of abscissas. We use the variable y to represent the real numbers on the vertical axis, which will now be known as the "y axis," or the axis of ordinates. Each point (x, y) on the plane corresponds to a unique ordered pair of real numbers. The point (x, y) on the plane is that point located by finding the real number x on the horizontal axis and then the point that is at the vertical distance from this point indicated by the location of the real number y on the vertical axis. For example, the point $(3, -2)$ is located 2 units below the point 3 on the x axis.

This scheme establishes a 1 : 1 correspondence between the points of the plane and the set K_2 of ordered pairs of real numbers. The plane, with the x axis and the y axis, is called the *cartesian plane*.

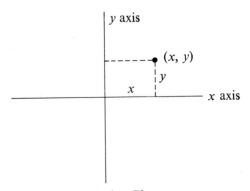

The Cartesian Plane

When the ordered pairs in a relation are "mapped" onto a plane, the result is called a *graph* of the relation. Of course, it is necessary to exaggerate when we construct the graph of a relation. The graph of a point is actually a small circle, or dot; the graph of a line, or a curve, must be made visible by using a solid mark, and so on.

4.5. EXAMPLES OF GRAPHS OF RELATIONS. We present here examples of seven relations together with their graphs. We emphasize that the graph of a relation should be sketched whenever we study the relation.

4.5.1. The graph of the relation $\{(0, 1), (1, 0)\}$ is

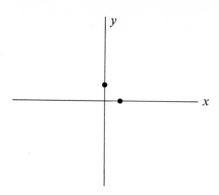

4.5.2. The graph of the relation $\{(x, y) ; x + y \leq 4, x, y \in N\}$ is

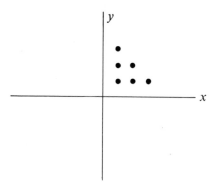

4.5.3. The graph of the relation

$$\{(x, y) ; 2x + y = 1, x^2 \geq 0, y^2 \geq 0, x \geq 1\}$$

is the "half" line h:

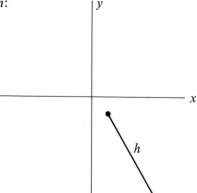

4.5.4. The graph of the relation

$$\{(x, y) \; ; \; x^2 + y^2 = 4, |x| \leq 2, |y| \leq 2\}$$

is the circle whose center is at $(0, 0)$ and whose radius is 2:

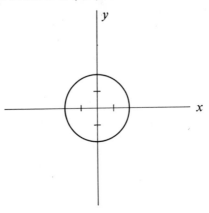

4.5.5. The graph of the relation

$$\{(x, y) \; ; \; x^2 + y^2 = 4, 0 \leq x \leq 2, |y| \leq 2\}$$

is the right semicircle with center at $(0, 0)$ and radius 2:

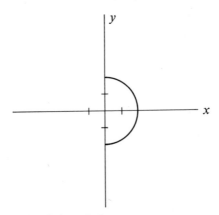

4.5.6. The graph of the relation

$$\{(x, y) \; ; \; x^2 + y^2 < 4, \; 0 < x < 2, \; 0 < y < 2\}$$

is the interior of the upper right quarter circle whose center is at $(0, 0)$ and radius is 2:

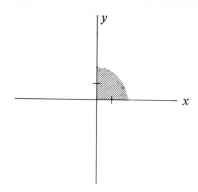

4.5.7.   The graph of the relation

$$\{(x, y) \,; y = |x|, x \in K\}$$

consists of the two half lines:

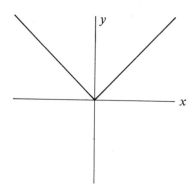

The two half lines come from the definition of the absolute value. The right half line is the set of points for which $y = |x| = x$ for $x \geq 0$; and the left half line comes from $y = |x| = -x$ for $x < 0$.

4.6. DOMAIN AND RANGE OF A RELATION.   The set of all first elements in a relation is called the *domain* of the relation. The set of all second elements in a relation is called the *range* of the relation.

In the examples of relations in the last section:

For 4.5.1 the domain is $\{0, 1\}$ and the range is $\{0, 1\}$.
For 4.5.2 the domain is $\{1, 2, 3\}$ and the range is $\{1, 2, 3\}$.

For 4.5.3 the domain is $\{x \; ; \; x \geq 1\}$ and the range is $\{y \; ; \; y \leq -1\}$.
For 4.5.4 the domain is $\{x \; ; \; |x| \leq 2\}$ and the range is $\{y \; ; \; |y| \leq 2\}$.
For 4.5.5 the domain is $0 \leq x \leq 2$ and the range is $\{y \; ; \; |y| \leq 2\}$.
For 4.5.6 the domain is $\{x \; ; \; 0 < x < 2\}$ and the range is

$$\{y \; ; \; 0 < y < 2\}.$$

For 4.5.7 the domain is $\{x \; ; \; x \in K\}$ and the range is $\{y \; ; \; y \geq 0\}$.

4.7. FUNCTION. A relation in which no two of its elements have the same first element is called a *function*. If two points, (x, y_1) and (x, y_2), belong to a function, then $y_1 = y_2$. It follows that a vertical line will cross the graph of a function no more than one time.

If we denote the domain of a given function by X and its range by Y, we see that:

a. for every $x \in X$ there corresponds, as a second element in the ordered pair of which x is the first element, one and only one $y \in Y$, and

b. each second element $y \in Y$ in an ordered pair of numbers belonging to the function may have more than one $x \in X$ as its first element.

It is customary to denote the element of the range which corresponds to a certain element $x \in X$ of the domain by the symbol $f(x)$, which is read "function of x," or, more briefly, "f of x." In a function, we may call the representative symbol $x \in X$, the *independent variable* and the representative symbol $y = f(x) \in Y$, the *dependent variable*. We shall use the symbol $f(x)$ for a representative element of the range only if we have a function, and shall avoid using this symbol when the relation is not a function.

In Section 4.5 only the relations 4.5.1, 4.5.3, and 4.5.7 are functions.

In working with relations and functions there must always be an understanding of precisely what the relation or function is meant to be, and the domain and range must be clearly defined. However, it may not be necessary to make the complete description in the terminology of the language of mathematics. In general, unless it is stated otherwise, the domain is understood to be either the set K, or some obvious subset of K. If the domain is obvious, or clearly defined, then it is customary to assume that the range is the subset of K that is determined by the defining statement and the given domain. For example, the open statement $x^2 + y^2 = 25$ is considered sufficient to describe the relation consisting of the points on the circle with center at $(0, 0)$ and radius 5, that is, all of the real values of x and y that will make the open statement true. If we write the open statement $y = 3x - 2$ we shall mean the function consisting of the points on the line on which the coordinates of each point make the open statement true. If we mean only a portion of the circle, or a segment of the line, then more is required in the definition of the relation or function.

PROBLEM SET 4.2

1. Identify each of the following as a relation or as a function. Determine the domain and the range, and sketch the graph.

   a. $\{(1, 2), (2, 1), (3, 2), (5, -1), (1, 3)\}$;

   b. $\{(x, y) ; y = 2x + 1, 0 \leq x \leq 2\}$;

   c. $\{(x, y) ; x + y = 1, x > 3\}$;

   d. $\{(x, y) ; x - 3y = 1 \text{ and } 3x + y = 1\}$;

   e. $\{(x, y) ; x - 3y = 1 \text{ or } 3x + y = 1\}$;

   f. $y = \sqrt{16 - x^2}$ (*N.B.* Remember that $\sqrt{k^2} = |k|$);

   g. $y = 3x^2 + 2 \text{ and } x \geq 0$;

   h. $x^2 + y^2 = 25 \text{ and } x \in \{1, 2, 3, 4\}$;

   i. $y = \begin{cases} 0, & \text{for } x \text{ rational and } 0 \leq x \leq 1, \\ 1, & \text{for } x \text{ irrational and } 0 \leq x \leq 1; \end{cases}$

   j. $y = \begin{cases} \sqrt{x} \text{ whenever } x \geq 0, \\ -\sqrt{-x} \text{ whenever } x < 0; \end{cases}$

   k. $y = x + |x|$;

   l. $y = \begin{cases} 0, & \text{if } x \geq 1, \\ 1, & \text{if } x < 1; \end{cases}$

   m. $y = |x - 2|$;

   n. $x^2 + y^2 = 36 \text{ and } -1 \leq x \leq 4$;

   o. $x^2 + y^2 = 36 \text{ and } -1 \leq y \leq 4$;

   p. $x^2 + y^2 < 25$;

q. $x^2 + y^2 > 25$ and $y \geq 0$;

r. $2x - 3y \geq 6$;

s. $3x + y = 7$ and $5x - 3y = 7$;

t. $3x + y > 7$ and $5x - 3y < 7$;

2.   Given that the symbol $[x]$ means the largest integer not greater than x, so that, for example, $[3/2] = 1$, $[-3/2] = -2$, for every x in the number interval $[0, 1)$, $[x] = 0$, for every x in the number interval $(-5, -4]$, $[x] = -5$, and so on. Study each of the following relations as you did those in Problem 1:

     a. $y = [x]$;

     b. $y = x + [x]$;

     c. $y = x - [x]$;

     d. $y = [3x]$;

     e. $y = 3x + [2x]$.

3.   Prove that the set of ordered pairs, $\{(p, q) \, ; p, q \in N\}$ is a countable infinite set.

4.   Prove that the set of ordered pairs of rational numbers,

$$\{(a, b) \, ; a, b \in R\},$$

is a countable infinite set.

CHAPTER FIVE

Rudimentary Algebras

5.1. OPERATION. If, in a set S of distinct elements, there is a rule, or a process, which prescribes a unique element for each ordered finite subset $P \subseteq S$ containing n elements, then, this process is called an *n-ary operation*. The unique element prescribed by this operation for each subset $P \subseteq S$ of n elements is called the *result*.

If $n = 1$, then P contains one element, and the operation is called a *unary* operation. In a unary operation, each element of S operates upon itself to produce a result. Examples of unary operations include squaring a number, raising a number to an integral power, taking an integral root of a number, and finding the factorial of a number.

If $n = 2$, then P contains two elements, and the operation is called a *binary* operation. With a binary operation, $\mathbf{o}$, every pair of elements $a, b \in S$ produces the result $(a \mathbf{o} b)$. Examples of binary operations include ordinary addition, ordinary multiplication, finding the least common multiple of two numbers, and finding the greatest common divisor of two numbers.

5.2. CLOSURE, IDENTITY ELEMENT, INVERSE ELEMENT. In most algebras we use a binary operation, so that we shall simply use the word "operation" to mean "binary operation." If other than a binary operation is to be used, we shall say so. We shall use the symbol "$\mathbf{o}$" to designate an undefined binary operation. A binary operation may be used, in general, for an element to operate on itself. Thus, in a statement which states "for every $a, b \in S$" the results $a \mathbf{o} a$ and $b \mathbf{o} b$ are meant to be included.

If, for every $a, b \in S$, the result $a \mathbf{o} b \in S$, then we say that S is *closed* with respect to the operation $\mathbf{o}$, or that S has *closure* with respect to the operation $\mathbf{o}$.

If there is an element $e \in S$, such that, for every $a \in S$,

$$a \circ e = a,$$

then, e is called an *identity element* with respect to the operation $\circ$. For example, for the operation of ordinary addition in the set K, of real numbers, the identity element is 0, since for every $k \in K$, $k + 0 = k$; similarly, for the operation of ordinary multiplication in K, the identity element is 1, since, for every $k \in K$, $k \times 1 = k$.

If, for an element $a \in S$, we can find an element a', which is not necessarily distinct from a, such that

$$a \circ a' = e,$$

then a' is called the *inverse* of a, or the *reciprocal* of a, or the *negative* of a, with respect to the operation $\circ$. For example: in the set K with the operation of ordinary addition the inverse of each $k \in K$ is $(-k)$, since $k + (-k) = 0$; and the inverse of $k \in K$ with respect to the operation of ordinary multiplication is $1/k$, providing $k \neq 0$, since $k \times 1/k = 1$. The inverse of $a \in S$ with respect to the operation $\circ$ may be designated by a', or $1/a$, or $-a$, or a^{-1}.

5.3. LAWS FOR BINARY OPERATIONS. Whenever we have a set S of distinct elements and one or two different binary operations, $\circ$ and $\#$, the following properties, or laws, should be tested.

5.3.1. *Closure.* $\forall \, a, b \in S, a \circ b \in S; a \# b \in S.$

5.3.2. *Commutative Law.* $a \circ b = b \circ a; a \# b = b \# a.$

5.3.3. *Associative Law.* $a \circ (b \circ c) = (a \circ b) \circ c;$
$a \# (b \# c) = (a \# b) \# c.$

5.3.4. *Distributive Laws.* $a \circ (b \# c) = (a \circ b) \# (a \circ c);$
$a \# (b \circ c) = (a \# b) \circ (a \# c).$

5.4. MODULAR OPERATIONS. For our first examples of rudimentary algebras we shall examine three separate algebras for which the set is a finite set and the operations are known as modular operations.

5.4.1. *Modulo 5 Algebra.* Given the finite set N_5,

$$N_5 = \{1, 2, 3, 4, 5\},$$

in which we define the two operations:

$+$ (mod 5) in which we first find the ordinary sum and then reduce the result by 5 or a multiple of 5 until the remainder is 5 or less than 5, for example, $2 + 4 = 1$ (mod 5) and $3 + 2 = 5$ (mod 5), and

$\times$ (mod 5) in which the ordinary product is obtained and then it is reduced by 5 or a multiple of 5 until the remainder is 5 or less than 5, for example, $2 \times 4 = 3$ (mod 5) and $3 \times 5 = 5$ (mod 5).

Since N_5 is a finite set we can put all of the results for each operation in a "multiplication" table. In each of the following tables the first factor is in the column at the left and the second factor is in the top row.

$+$ (mod 5)	1	2	3	4	5
1	2	3	4	5	1
2	3	4	5	1	2
3	4	5	1	2	3
4	5	1	2	3	4
5	1	2	3	4	5

$\times$ (mod 5)	1	2	3	4	5
1	1	2	3	4	5
2	2	4	1	3	5
3	3	1	4	2	5
4	4	3	2	1	5
5	5	5	5	5	5

The Multiplication Tables for Modulo 5 Algebra

Since each multiplication table contains all possible pairs of factors from N_5, and since every result for each operation is an element of N_5, it follows that N_5 is closed with respect to each of the two operations.

We find from the first table that 5 is the identity element for the operation of $+$ (mod 5). The second table tells us that 1 is the identity element for $\times$ (mod 5).

If we use the notation that $-a$ is the reciprocal of a with respect to the additive operation, $+ \pmod 5$, we find from the first table that

$$-4 = 1, \quad -1 = 4, \quad -2 = 3, \quad -3 = 2, \quad \text{and} \quad -5 = 5 \pmod 5.$$

Similarly, if we use the notation that "$1/a$" is the multiplicative reciprocal of a with respect to the operation $\times \pmod 5$, we find from the second table that

$$1/2 = 3, \quad 1/3 = 2, \quad 1/4 = 4, \quad \text{and } 1/5 \text{ does not exist} \pmod 5.$$

We may solve open statements in modulo 5 algebra, as is demonstrated in the following examples:

a. $2x + 1 = 4 \pmod 5$ has the solution set $\{4\}$.
b. $2x + 4 = 3 \pmod 5$ has the solution set $\{2\}$.
c. $x^2 - 4 = x^2 + 1 = 5 \pmod 5$ has the solution set $\{2, 3\}$.
d. $x^2 - 1 = x^2 + 4 = 5 \pmod 5$ has the solution set $\{1, 4\}$.
e. $x^2 + x + 3 = 5 \pmod 5$ is solved by factoring in which the equation is written in the factored form $(x + 2)(x + 4) = 5 \pmod 5$. When we set each factor in turn equal to 5, we obtain the solution set $\{1, 3\}$.

5.4.2. *Modulo 2 Algebra.* Given the set $\{0, 1\}$ with the operations of $+ \pmod 2$ and $\times \pmod 2$, which are defined similarly to the way the operations for modulo 5 algebra were defined. The multiplication tables for modulo 2 algebra are

$+ \pmod 2$	0	1
0	0	1
1	1	0

$\times \pmod 2$	0	1
0	0	0
1	0	1

The Multiplication Tables for Modulo 2 Algebra

It is obvious that the closure property is valid. The identity element for $+ \pmod 2$ is 0, and the identity element for $\times \pmod 2$ is 1. The element 1 is its own reciprocal for each operation. The element 0 has no inverse with respect to $\times \pmod 2$.

Examples of solution sets for open statements are:

a. $x + 1 = 0$ (mod 2) has the solution set $\{1\}$.
b. $x^2 + 1 = 0$ (mod 2) has the solution set $\{1\}$.
c. $x^2 - x + 1 = 0$ (mod 2) has the solution set $\varnothing$.

5.4.3. *Modulo 10 Algebra.* Given the set

$$D = \{0, 1, 2, 3, 4, 5, 6, 7, 8, 9\}$$

and the operations $+$ (mod 10) and $\times$ (mod 10). These operations are defined similarly to the definition of the operations for modulo 5 algebra. It should be observed that the result for either of the modulo 10 operations is accomplished by first performing the usual addition, or multiplication, and then retaining only the "units digit" in the result. The "multiplication" tables for modulo 10 algebra are given here.

+ (mod 10)	0	1	2	3	4	5	6	7	8	9
0	0	1	2	3	4	5	6	7	8	9
1	1	2	3	4	5	6	7	8	9	0
2	2	3	4	5	6	7	8	9	0	1
3	3	4	5	6	7	8	9	0	1	2
4	4	5	6	7	8	9	0	1	2	3
5	5	6	7	8	9	0	1	2	3	4
6	6	7	8	9	0	1	2	3	4	5
7	7	8	9	0	1	2	3	4	5	6
8	8	9	0	1	2	3	4	5	6	7
9	9	0	1	2	3	4	5	6	7	8

The Table for $+$ (mod 10)

It is immediately evident that D is closed for $+$ (mod 10). The identity element is 0. The "negative" numbers in D are

$$-1 = 9, \quad -9 = 1, \quad -2 = 8, \quad -8 = 2, \quad -3 = 7, \quad -7 = 3,$$
$$-4 = 6, \quad -6 = 4, \quad -5 = 5 \text{ (mod 10)}.$$

× (mod 10)	0	1	2	3	4	5	6	7	8	9
0	0	0	0	0	0	0	0	0	0	0
1	0	1	2	3	4	5	6	7	8	9
2	0	2	4	6	8	0	2	4	6	8
3	0	3	6	9	2	5	8	1	4	7
4	0	4	8	2	6	0	4	8	2	6
5	0	5	0	5	0	5	0	5	0	5
6	0	6	2	8	4	0	6	2	8	4
7	0	7	4	1	8	5	2	9	6	3
8	0	8	6	4	2	0	8	6	4	2
9	0	9	8	7	6	5	4	3	2	1

The Table for × (mod 10)

The set D is closed for × (mod 10). The identity element is 1. The reciprocals are

$$1/1 = 1, \quad 1/3 = 7, \quad 1/7 = 3, \quad 1/9 = 9 \,(\text{mod } 10).$$

The numbers 0, 2, 4, 5, 6, 8 have no inverses for × (mod 10).

The product

$$x \times y = 0 \,(\text{mod } 10)$$

requires either that one of the factors, x or y, be 0, or that one of the factors is even and the other is 5.

Since each of the numbers in D can be "factored" into the product of two other numbers of D, in at least one way, there are no prime numbers in D. For example, $3 = 1 \times 3 = 7 \times 9$, and $7 = 1 \times 7 = 3 \times 9$. If we use the usual meaning for exponents: $a^2 = a \times a \,(\text{mod } 10)$, then 5 is an idempotent element, since $5^2 = 5 \,(\text{mod } 10)$.

The "square" numbers in D are 0, 1, 4, 5, 6, 9, and we have

$$1^2 = 9^2 = 1, \quad 2^2 = 8^2 = 4, \quad 3^2 = 7^2 = 9, \quad 4^2 = 6^2 = 6 \,(\text{mod } 10).$$

If we define a "fraction" a/b to be the number that produces the result a when it is multiplied by b, that is,

$$a/b \times b = a \,(\text{mod } 10),$$

then algebra modulo 10 contains certain fractions and does not contain certain others. For some examples,

1/2 does not exist, but $2/4 = 3$ or $2/4 = 8$, $4/8 = 3$ or $4/8 = 8$, 3/6 does not exist, $5/1 = 5$, $6/2 = 3$ or 8, 7/4 does not exist, $8/6 = 3$ or 8, 9/8 does not exist.

$x/5 = 0$ if x is even, but $x/5 = 5$ if x is odd.

$0/x = 0$ if x is odd but $0/x = 0$ or 5 if x is even.

An open statement with the form $ax = b \,(\text{mod } 10)$ has no solution if a is even and b is odd, or if $a = 5$ and $b \neq 0$ or $b \neq 5$. The solution set for $2x = 4 \,(\text{mod } 10)$ is $\{2, 7\}$. The solution set for $3x = 7$ is $\{9\}$.

An open statement with the form $x^2 = a$ has no solution if $a = 2$, 3, 7, or 8. The solution set for $x^2 = 1$ is $\{1, 9\}$.

A quadratic equation

$$x^2 - 5x + 6 = x^2 + 5x + 6 = 0$$

may be solved by factoring in a procedure similar to that used in ordinary algebra. The factored form of the equation is

$$(x + 2)(x + 3) = 0.$$

The product of two factors is 0 if one of the factors is 0. If each of the factors is, in turn, set equal to 0, we obtain the solution set $\{8, 7\}$. But, the product of two factors may be 0 if one factor is 5 and the other factor is even, so we set $(x + 2) = 5$ to obtain $x = 3$. Since $(x + 3)$ is even, equal to 6, when $x = 3$, we have obtained another root, $x = 3$. On the other hand, we may also set the other factor, $(x + 3)$, equal to 5 to obtain the root $x = 2$. Since $x = 2$ makes $(x + 2)$ even, the fourth root of the given equation is $x = 2$. The complete solution set for $x^2 + 5x + 6 = 0$ is $\{2, 3, 7, 8\}$.

The quadratic equation

$$x^2 + 4 = 0$$

has the factored form

$$(x + 4)(x + 6) = 0.$$

The complete solution set is obtained from setting each factor in turn equal to zero. If either factor is equal to 5, the value for x will make the other factor odd, so that the product of the two factors will be 5 instead of 0. The solution set is $\{4, 6\}$.

The cubes of the numbers in D are:

$$1^3 = 1, \quad 2^3 = 8, \quad 3^3 = 7, \quad 4^3 = 4, \quad 5^3 = 5, \quad 6^3 = 6, \quad 7^3 = 3,$$
$$8^3 = 2, \quad 9^3 = 9, \text{(mod 10)}.$$

5.5. ALGEBRAS WITH SETS OF ORDERED PAIRS OF NUMBERS.

We shall demonstrate three examples of rudimentary algebras in which the elements are sets of ordered pairs of numbers. In each algebra we shall define (1) equality of one element to another, (2) multiplication of an element by a "scalar" k, where "scalar" means a single real number, or a constant, (3) addition of two of the ordered pair elements, and (4) multiplication of two of the ordered pair elements. We shall then determine the identity and inverse elements for each of addition and multiplication. Finally, we shall solve a few examples of open statements in the algebra being examined.

5.5.1. An Algebra in I_2.

Given the set

$$I_2 = I \times I = \{(a, b) \; ; a, b \in I\},$$

the set of ordered pairs of integers. We establish the following definitions, in which the operations $a + b$, ab, where a and b are integers, indicate the result of ordinary addition and ordinary multiplication of integers:

5.5.1.1. *Equality*: $(a, b) = (c, d) \leftrightarrow ad = bc$.

5.5.1.2. *Multiplication by a scalar:* $k(a, b) = (ka, b)$.

5.5.1.3. *Addition:* $(a, b) + (c, d) = (ad + bc, bd)$.

5.5.1.4. *Multiplication:* $(a, b) \times (c, d) = (ac, bd)$.

For every $k \in K$, $k \neq 0$, the definition of equality provides that

$$(ka, kb) = (a, b),$$

since $kab = kab$.

The identity element for addition is $(0, b) = (0, 1)$, $b \neq 0$, because

$$(c, d) + (0, b) = (bc, bd) = (c, d), \quad \text{or}$$
$$(c, d) + (0, 1) = (c, d).$$

The identity element for multiplication is $(b, b) = (1, 1)$, $b \neq 0$, because

$$(c, d) \times (b, b) = (bc, bd) = (c, d), \quad \text{or}$$
$$(c, d) \times (1, 1) = (c, d).$$

The inverse of (c, d) with respect to addition is $(-c, d)$ since

$$(c, d) + (-c, b) = (0, d^2) = (0, 1).$$

The inverse of (c, d) with respect to multiplication is (d, c) since

$$(c, d) \times (d, c) = (cd, cd) = (1, 1).$$

We may use the "exponent" notation, $(x, y)^2$, to mean $(x, y) \times (x, y) = (x^2, y^2)$. Numerical coefficients are scalars, so, for example, $k(x, y) = (kx, y)$.

We may find solution sets for open statements, as is demonstrated in the following examples:

(a) $(3, 2) \times (x, y) = (5, 3)$ may be solved by first multiplying the factors in the left member to obtain

$$(3x, 2y) = (5, 3),$$

which, from the definition of equality, requires that

$$9x = 10y.$$

This equation is satisfied if $x = 10k$ and $y = 9k$, so that the required solution set is

$$\{(10k, 9k) \, ; k \in K\}.$$

(b) Solve the open statement

$$[4(1, -3) \times (x, y)] + (5, -4) = (7, -1).$$

The first term of the left member reduces to

$$4(1, -3) \times (x, y) = (4, -3) \times (x, y) = (4x, -3y),$$

so the open statement has become

$$(4x, -3y) + (5, -4) = (7, -1),$$

and the left member of the latter open statement reduces to

$$(-16x - 15y, 12y) = (7, -1).$$

This reduced equality is possible, according to the definition of equality, if

$$16x + 15y = 84y, \quad \text{or} \quad 16x = 69y.$$

The solution set that satisfies the latter condition is

$$\{(69k, 16k) \, ; \, k \in K, k \neq 0\}.$$

(c)   Find the solution set for

$$(x, y)^2 - 5(x, y) = -6(1, 1).$$

The left member reduces to

$$(x^2y - 5xy^2, y^3) = (x^2 - 5xy, y^2),$$

so the open statement becomes

$$(x^2 - 5xy, y^2) = (-6, 1).$$

From the definition of equality this is possible if

$$x^2 - 5xy = -6y^2,$$

which reduces by ordinary algebraic methods to

$$(x - 2y)(x - 3y) = 0.$$

Consequently, the solution set contains the union of two sets of ordered pairs of integers:

$$\{(2k, k) \text{ and } (3k, k) \, ; \, k \in K, k \neq 0\}.$$

Since, for every $k \in K$, $k \neq 0$, $(2k, k) = (2, 1)$ and $(3k, k) = (3, 1)$ we shall call $(2, 1)$ and $(3, 1)$ the *principal solutions*.

5.5.2. *A Rudimentary Algebra in R_2.*   We are given the set or ordered pairs of rational numbers,

$$R_2 = R \times R = \{(p, q) \, ; \, p, q \in R\}.$$

A rudimentary algebra in R_2 is established by the following definitions:

5.5.2.1. *Equality:* $(p, q) = (r, s) \leftrightarrow p = r$ and $q = s$.

5.5.2.2. *Multiplication by a scalar:* $k(p, q) = (kp, kq)$.

5.5.2.3. *Addition:* $(p, q) + (r, s) = (p + r, q + s)$.

5.5.2.4. *Multiplication:* $(p, q) \times (r, s) = (pr + 2qs, ps + qr)$.

The algebra is obviously closed since the result for addition and for multiplication is an ordered pair of rational numbers.

The identity with respect to addition is $(0, 0)$, since

$$(p, q) + (0, 0) = (p, q).$$

The identity with respect to multiplication is $(1, 0)$, since

$$(p, q) \times (1, 0) = (p, q).$$

The additive inverse of (p, q) is $(-p, -q)$, since

$$(p, q) + (-p, -q) = (0, 0).$$

The multiplicative inverse of (p, q) is $(p/v, -q/v)$, where

$$v = p^2 - 2q^2 \neq 0,$$

since

$$(p, q) \times (p/v, -q/v) = (p^2/v - 2q^2/v, 0) = (1, 0).$$

If $(x, y)^2 = (x, y) \times (x, y)$ we have

$$(x, y)^2 = (x^2 + 2y^2, 2xy).$$

For examples of open statements in this algebra, we examine the following:

(a) $(3, -5) \times (x, y) = (-4, 2)$

is solved by first completing the operation in the left member to obtain

$$(3x - 10y, -5x + 3y) = (-4, 2).$$

Then, the definition of equality gives us

$$3x - 10y = -4, \quad \text{and}$$
$$-5x + 3y = 2.$$

When this set of simultaneous linear equations is solved by methods of ordinary algebra we find that the ordered pair of rational numbers

$$(8/41, -14/41)$$

is the solution. This is also the solution of our original open statement.

(b) The solution of

$$(x, y)^2 = (22, 12)$$

is accomplished by expanding the left member to obtain

$$(x, y)^2 = (x^2 + 2y^2, 2xy) = (22, 12).$$

Then, the definition of equality requires that

$$x^2 + 2y^2 = 22, \quad \text{and}$$
$$2xy = 12.$$

When the values of x and y which simultaneously satisfy these two equations are obtained, we find that the solution of the open statement is $(2, 3)$, and $(-2, -3)$.

5.5.3. *A Second Rudimentary Algebra in* I_2. We may, of course, define different operations in the same set to form different algebras. In this rudimentary algebra we have the set of ordered pairs of integers, I_2, and the following definitions:

5.5.3.1. *Equality:* $(a, b) = (c, d) \leftrightarrow a = c$ and $b = d$.
5.5.3.2. *Multiplication by a scalar:* $k(a, b) = (ka, kb)$.
5.5.3.3. *Addition:* $(a, b) + (c, d) = (a + c, b + d)$.
5.5.3.4. *Multiplication:* $(a, b) \times (c, d) = (b - d, c - a)$.

The identity element with respect to addition is $(0, 0)$, since

$$(a, b) + (0, 0) = (a, b).$$

There is no multiplicative identity. However, to each element (a, b) there is an element, $(b + a, b - a)$, such that

$$(a, b) \times (b + a, b - a) = (a, b).$$

Consequently, there are no multiplicative inverses.

The inverse of (a, b) with respect to addition is $(-a, -b)$ since

$$(a, b) + (-a, -b) = (0, 0).$$

We have the following interesting results in this algebra:

a. $(x, y)^2 = (x, y) \times (x, y) = (0, 0)$;
b. $(x, y) \times (0, 0) = (y, -x)$;
c. $(c, d) \times (a, b) = (d - b, a - c) = -1(b - d, c - a)$,

so multiplication is not commutative.

d. $(a, b) \times [(c, d) \times (e, f)] = (a, b) \times (d - f, e - c)$
$$= (b - e + c, d - f - a);$$
$[(a, b) \times (c, d)] \times (e, f) = (b - d, c - a) \times (e, f)$
$$= (c - a - f, e - b + d),$$

so multiplication is not associative.

e. $(a, b) \times [(c, d) + (e, f)] = (a, b) \times (c + e, d + f)$
$$= (b - d - f, c + e - a),$$
$[(a, b) \times (c, d)] + [(a, b) \times (e, f)] = (b - d, c - a)$
$$+ (b - f, e - a) = (2b - d - f, c + e - 2a),$$

so multiplication is not distributive over addition.

f. $(a, b) + [(c, d) \times (e, f)] = (a, b) + (d - f, e - c)$
$$= (a + d - f, b + e - c),$$
$[(a, b) + (c, d)] \times [(a, b) + (e, f)] = (a + c, b + d)$
$$\times (a + e, b + f) = (d - f, e - c),$$

so addition is not distributive over multiplication.

5.6. AN ALGEBRA IN K_3. Given the set of ordered triples of real numbers

$$K_3 = K \times K \times K = \{(x, y, z) \; ; x, y, z \in K\},$$

in which we establish the following definitions for a rudimentary algebra:

5.6.1. *Equality.* $(a_1, a_2, a_3) = (b_1, b_2, b_3) \leftrightarrow a_1 = b_1, a_2 = b_2$ and $a_3 = b_3$.

5.6.2. *Multiplication by a Scalar.* $k(a_1, a_2, a_3) = (ka_1, ka_2, ka_3)$.

5.6.3. *Addition.* $(a_1, a_2, a_3) + (b_1, b_2, b_3)$
$= (a_1 + b_1, a_2 + b_2, a_3 + b_3)$.

5.6.4. *The Dot Product.* $(a_1, a_2, a_3) \cdot (b_1, b_2, b_3) = a_1b_1 + a_2b_2 + a_3b_3$.

5.6.5. *The Cross Product.* $(a_1, a_2, a_3) \times (b_1, b_2, b_3)$
$= (a_2b_3 - a_3b_2, -a_1b_3 + a_3b_1, a_1b_2 - a_2b_1)$.

In this algebra we have defined three operations. While K_3 is closed with respect to addition and the cross product, it is not closed with respect to the dot product, since the result in the dot product is a single real number instead of an ordered triple of real numbers. The cross product is much easier to calculate if we first introduce the idea of a determinant of order two.

A determinant of order two is a square array of real numbers in two rows and two columns. We designate the four real numbers in the array by the symbols $a_{11}, a_{12}, a_{21}, a_{22}$ in which we use the double subscripts to locate the number in the array. The first digit in the subscript is the row address and the second digit is the column address; thus, the address of the element a_{11} is the first row and first column, the address of a_{12} is the first row and second column, that of a_{21} is the second row and first column, and the address of a_{22} is the second row and second column. The complete definition of the determinant of order two is that it is a square array of numbers with a definite numerical value. The determinant of order two and its numerical value are

$$\begin{vmatrix} a_{11} & a_{12} \\ a_{21} & a_{22} \end{vmatrix} = a_{11}a_{22} - a_{21}a_{12}.$$

For an example, the determinant of order two

$$\begin{vmatrix} 4 & 3 \\ 6 & 5 \end{vmatrix} = 20 - 18 = 2.$$

We return to the definition of the cross product. If the first factor is written in the first row, and the second factor is written in the second row, of an array of two rows and three columns, we obtain, for the cross product $(a_1, a_2, a_3) \times (b_1, b_2, b_3)$, the array

$$
\begin{array}{ccc}
a_1 & a_2 & a_3 \\
b_1 & b_2 & b_3.
\end{array}
$$

From this array we can form three determinants of order two by successively omitting one column at a time. The first element in the cross product is the determinant of order two obtained by omitting the first column, the second element is the negative of the determinant of order two obtained by omitting the second column, and the third element is obtained by omitting the third column in the array. Thus, we have

$$
(a_1, a_2, a_3) \times (b_1, b_2, b_3) = \left(\begin{vmatrix} a_2 & a_3 \\ b_2 & b_3 \end{vmatrix}, \; -\begin{vmatrix} a_1 & a_3 \\ b_1 & b_3 \end{vmatrix}, \; \begin{vmatrix} a_1 & a_2 \\ b_1 & b_2 \end{vmatrix} \right).
$$

For example,

$$
(3, 4, -1) \times (4, -2, 3) = \begin{array}{ccc} 3 & 4 & -1 \\ 4 & -2 & 3 \end{array} = (10, -13, -22).
$$

It is obvious that addition is commutative and associative. The dot product is commutative, but associativity has no meaning in the dot product. The cross product is not commutative, since interchanging the factors in the cross product would change the sign in each element, so that

$$
(b_1, b_2, b_3) \times (a_1, a_2, a_3) = -1[(a_1, a_2, a_3) \times (b_1, b_2, b_3)].
$$

The associative law does not hold for the cross product, as the reader may easily verify. The dot product is distributive over addition, and the cross product is distributive over addition.

The solution of open statements will be discussed when we examine this algebra in greater detail as "vector algebra" in Chapter Eight.

5.7. THE ALGEBRA OF 2×2 MATRICES. A rectangular array of real numbers, consisting of m horizontal rows and n vertical columns, is called an $m \times n$ *matrix*. We are interested, in this section, in the set of 2×2 matrices. We shall use the same double subscript address for the elements in a 2 × 2 matrix as we used for the elements in a 2 × 2 determinant. In a 2 × 2 matrix we do not have a single real number to represent the numerical value, since the matrix is defined as simply an array of numbers. A matrix is distinguished from a determinant by enclosing the matrix inside curved lines, or large parentheses, and by enclosing the determinant inside vertical lines. We shall use capital letters to designate an entire matrix. Thus, the matrix A is

$$A = \begin{pmatrix} a_{11} & a_{12} \\ a_{21} & a_{22} \end{pmatrix}, \qquad B = \begin{pmatrix} b_{11} & b_{12} \\ b_{21} & b_{22} \end{pmatrix}$$

and the matrix B will designate the matrix containing the elements $b_{11}, b_{12}, b_{21}, b_{22}$. Thus, the set of all 2 × 2 matrices is

$$\left\{ \begin{pmatrix} x_{11} & x_{12} \\ x_{21} & x_{22} \end{pmatrix} ; x_{11}, x_{12}, x_{21}, x_{22} \in K \right\}.$$

We establish the algebra of 2 × 2 matrices with the following definitions:

5.7.1. *Equality.* $A = B \leftrightarrow a_{11} = b_{11}, a_{12} = b_{12}, a_{21} = b_{21}$, and $a_{22} = b_{22}$.

5.7.2. *Multiplication by a Scalar.*

$$kA = \begin{pmatrix} ka_{11} & ka_{12} \\ ka_{21} & ka_{22} \end{pmatrix}.$$

5.7.3. *Matrix Addition.*

$$A + B = \begin{pmatrix} (a_{11} + b_{11}) & (a_{12} + b_{12}) \\ (a_{21} + b_{21}) & (a_{22} + b_{22}) \end{pmatrix}.$$

5.7.4. *Matrix Product.*

$$AB = \begin{pmatrix} (a_{11}b_{11} + a_{12}b_{21}) & (a_{11}b_{12} + a_{12}b_{22}) \\ (a_{21}b_{11} + a_{22}b_{21}) & (a_{21}b_{12} + a_{22}b_{22}) \end{pmatrix}.$$

The matrix product is defined as follows:

If $AB = C$, then

c_{11} = the dot product of the elements in the first row of A with the elements in the first column of B;

c_{12} = the dot product of the elements in the first row of A with the elements in the second column of B;

c_{21} = the dot product of the elements in the second row of A with the elements in the first column of B;

c_{22} = the dot product of the elements in the second row of A with the elements in the second column of B.

For example,

$$\begin{pmatrix} 3 & -1 \\ 2 & 5 \end{pmatrix} \times \begin{pmatrix} p & q \\ r & s \end{pmatrix} = \begin{pmatrix} (3p - r) & (3q - s) \\ (2p + 5r) & (2q + 5s) \end{pmatrix}.$$

Since the result with either addition or multiplication is a 2×2 matrix with real numbers as elements, this algebra is closed with respect to each operation. Matrix addition is both commutative and associative. Matrix multiplication is associative, but it is not commutative. Matrix product is distributive over matrix addition.

The identity with respect to matrix addition is the "zero" matrix in which every element is 0. This "zero" matrix will be designated by the symbol O. The identity matrix with respect to matrix product is the matrix I in which the element in the first row and first column, and the element in the second row and second column, is 1, while the other two elements are each 0. The two identity matrices are

$$O = \begin{pmatrix} 0 & 0 \\ 0 & 0 \end{pmatrix} \quad \text{and} \quad I = \begin{pmatrix} 1 & 0 \\ 0 & 1 \end{pmatrix}.$$

The inverse for A with respect to matrix addition is $-A$, in which every element is the negative of its corresponding element in A. The inverse for A with respect to matrix multiplication is the matrix A^{-1}, which is obtained from the matrix A by (1) interchanging the elements a_{11} and a_{22}, (2) negating the elements a_{12} and a_{21}, and (3) dividing each element in the resulting matrix by the value of the determinant having the same elements as the matrix A. The two inverse matrices are

$$-A = \begin{pmatrix} -a_{11} & -a_{12} \\ -a_{21} & -a_{22} \end{pmatrix}, \qquad A^{-1} = \frac{1}{|A|} \begin{pmatrix} a_{22} & -a_{12} \\ -a_{21} & a_{11} \end{pmatrix},$$

where $|A| = a_{11}a_{22} - a_{12}a_{21}$. It follows at once that if $|A| = 0$, then the matrix A has no inverse with respect to matrix product.

For an example, if

$$A = \begin{pmatrix} 3 & 2 \\ 5 & 4 \end{pmatrix}, \quad \text{then} \quad A^{-1} = \begin{pmatrix} 2 & -1 \\ -5/2 & 3/2 \end{pmatrix}$$

and we find that $AA^{-1} = I$.

Open statements involving matrices may be solved by using methods similar to those used in other algebras. In general, we have the following procedures:

a. A matrix equation with the form

$$X + A = B$$

will have a matrix solution with the form

$$X = B + (-A) = B - A.$$

b. A matrix equation with the form

$$AX + B = C, \quad \text{where} \quad |A| \neq 0,$$

can be solved in matrices in the following steps:

(1) $AX = C - B$,
(2) $A^{-1}(AX) = A^{-1}(C - B)$,
(3) since $A^{-1}(AX) = (A^{-1}A)X = IX = X$, we have
      $X = A^{-1}(C - B)$.

5.8. GROUP. We are now ready to construct algebras with more definite structure than those we have discussed so far, or, at least, we shall pay more attention to the structure than we have in the rudimentary algebras defined above. The first, and the simplest, structure in algebra which we shall define is called a *group*.

Given a set of distinct elements

$$G = \{a, b, c, \ldots\},$$

where G may be either a finite set or an infinite set. In G we have a binary operation $\mathbf{o}$ for which the following four axioms are satisfied:

5.8.1. *Closure.* $\forall\, a, b \in G, (a \mathbf{o} b) \in G$.

5.8.2. *Associativity.* $\forall\, a, b, c \in G, a \mathbf{o} (b \mathbf{o} c) = (a \mathbf{o} b) \mathbf{o} c$.

5.8.3. *Identity.* $\exists\, e \in G, \ni \forall\, a \in G, a \mathbf{o} e = e \mathbf{o} a = a$.

5.8.4. *Inverse Elements.* $\forall\, a \in G \exists\, a' \in G \ni a \mathbf{o} a'$
$= a' \mathbf{o} a = e$.

If all four of these axioms are satisfied, then we have a *group*. While it is not necessary that the commutative law be satisfied we may have the following additional axiom:

5.8.5. *Commutativity.* $\forall\, a, b \in G, a \mathbf{o} b = b \mathbf{o} a$.
If Axiom 5.8.5 is satisfied in addition to those which define the group, then the group is called a *commutative group,* or it is called an *abelian group,* after the Norwegian mathematician Niels Abel (1802–1829).

We shall designate a group consisting of the set G and the operation $\mathbf{o}$ by the symbol $G(\mathbf{o})$. If the set G is finite, then $G(\mathbf{o})$ is called a *finite group.* If G is an infinite set, then $G(\mathbf{o})$ is called an *infinite group.*

If there is a subset $G_1 \subseteq G$, and we have the groups $G_1(\mathbf{o})$ and $G(\mathbf{o})$ for the same operation $\mathbf{o}$, then $G_1(\mathbf{o})$ is a *subgroup* of $G(\mathbf{o})$.

Some examples of groups are:

a. In 5.4.1 the set N_5 with the operation $+ \pmod 5$ forms an abelian group. The number of elements in the set for a finite group is the *order* of the group. Consequently, the group $N_5(+ \pmod 5)$ is an abelian group of order five.

b. Given the set $C_4 = \{+1, -1, +i, -i\}$, where i is the imaginary unit so that $i^2 = -1$, with the operation of ordinary multiplication, then, $C_4(x)$ is a finite abelian group of order 4. The identity element is $+1$, the inverse of i is $-i$, the inverse of -1 is -1. The group $C_2(x)$, where $C_2 = \{+1, -1\}$, is a subgroup of order 2 of $C_4(x)$. It is interesting to note that the elements of $C_4(x)$ can be "generated" by powers of i, that is, since $i^2 = -1$, $i^3 = -i$, $i^4 = 1$, then

$$C_4 = \{i, i^2, i^3, i^4 = 1\}.$$

A group in which one element may generate all of the other elements by successive application of the group operation is called a *cyclic* group.

c.   Given the set

$$C_3 = \{1, w_1, w_2\},$$

where

$$w_1 = \frac{-1 + i\sqrt{3}}{2} \quad \text{and} \quad w_2 = \frac{-1 - i\sqrt{3}}{2}.$$

The operation is ordinary multiplication, so the reader may verify by using ordinary algebraic techniques that

$$w_1{}^2 = w_2, \quad w_2{}^2 = w_1, \quad w_1 w_2 = w_1{}^3 = w_2{}^3 = 1.$$

It follows that $C_3(x)$ is a cyclic abelian group of order 3. The generating element may be either w_1 or w_2. The identity element is 1. The inverse of w_1 is w_2, and conversely.

PROBLEM SET 5.1

1.   Make a complete study of each of the following rudimentary algebras, similar to the study of the algebra modulo 5:
   a.   the algebra modulo 6 for the set $\{0, 1, 2, 3, 4, 5\}$;
   b.   the algebra modulo 7 for the set $\{0, 1, 2, 3, 4, 5, 6\}$.

2.   Set up the multiplication table and test to see if this algebra is a group. Given a real number t, where $t \neq 0$ and $t \neq 1$, and the following set of six elements defined in terms of t,

$$S = \left\{ t, \frac{1}{t}, (1 - t), \frac{1}{(1 - t)}, \frac{(t - 1)}{t}, \frac{t}{(t - 1)} \right\},$$

we define the operation $\#$ in S as follows:

$\forall \, a, b \in S, \ a \# b = $ the result of substituting the second factor in the place of each t in the first factor.

For example,

$$\frac{1}{t} \# (1 - t) \doteq \frac{1}{(1 - t)}, \quad \text{and}$$

$$(1 - t) \# \frac{1}{t} = 1 - \frac{1}{t} = \frac{(t - 1)}{t}.$$

3. In the algebra in I_2, discussed in 5.5.1, compute each of the following results:

   a. $(5, 6) + (2, -14)$; c. $(3, 2) \times (4, 9)$;
   b. $(5, 6) \times (6, 5)$; d. $(0, 5) + (6, -9)$.

4. In the rudimentary algebra described in 5.5.2 compute each of the following results:

   a. $(a, b) \times (a, -b)$; d. $(a, b) \times (a, b)$;
   b. $(a, b) \times (1, 0)$; e. $(3/5, 4/5) \times (-3/5, 4/5)$;
   c. $(0, 1) \times (0, 1)$; f. $(2, 3) \times (2/13, -3/13)$.

5. In the algebra in I_2 discussed in 5.5.1, and the following definition of an "order" relation:

   5.5.1.5. *Order:* $(a, b) < (c, d) \leftrightarrow ad < bc$.

   Prove that if $(a, b) < (c, d)$, then $(a, b) < (a + c, b + d) < (c, d)$.

6. Compute the matrix products:

   a. $\begin{pmatrix} 3 & 2 \\ 1 & 5 \end{pmatrix} \times \begin{pmatrix} 4 & 5 \\ 7 & 9 \end{pmatrix}$; b. $\begin{pmatrix} 4 & -9 \\ 2 & -3 \end{pmatrix} \times \begin{pmatrix} 2 & 0 \\ 1 & 2 \end{pmatrix}$.

7. Given the following matrices:

   $$A = \begin{pmatrix} 1 & 0 \\ -1 & 0 \end{pmatrix}; \quad B = \begin{pmatrix} 1 & 1 \\ -1 & 1 \end{pmatrix}; \quad C = \begin{pmatrix} 1 & 1 \\ 0 & 1 \end{pmatrix}.$$

   Calculate each of the following results:

   a. $A + B$; d. $(A \times B) \times C$;
   b. $A \times B$; e. $A + (B \times C)$;
   c. $A \times (B \times C)$; f. $(A + B) \times (A + C)$.

8. In the algebra defined in K_3, which was discussed in Section 5.6, calculate each of the following:

    a. $(1, 4, 2) \times (3, 12, 6)$;   b. $(4, -2, 5) \times (7, 1, -3)$;

    c. $(3, 4, 2) \cdot (6, 5, -2)$;

    d.   If $\bar{i} = (1, 0, 0)$, $\bar{j} = (0, 1, 0)$, $\bar{k} = (0, 0, 1)$, calculate each of the following: $\bar{i} \times \bar{j}, \bar{j} \times \bar{k}, \bar{k} \times \bar{i}, \bar{i} \cdot \bar{j}, \bar{j} \cdot \bar{k}, \bar{k} \cdot \bar{i}$.

9.   Discuss each of the following rudimentary algebras defined over the set I of integers and the operations # and & defined as indicated in each case:

    a. $a \# b = (a + 2b)$ and $a \,\&\, b = 2ab$.

    b. $a \# b = (a + b^2)$ and $a \,\&\, b = ab^2$.

    c. $a \# b = a^b$ and $a \,\&\, b = b$.

10.   Test each of the following algebras to see if a group is defined. If it is not a group explain why it is not.

    a.   The set E of even natural numbers with ordinary addition;

    b.   The set $\{a + b\sqrt{2} \,; a, b \in R\}$ with ordinary multiplication.

    c.   The set $\{p, p^2, p^3, \ldots, p^n = 1\}$ with ordinary multiplication.

    d.   The set R of rational numbers with ordinary multiplication.

    e.   The set $\{a + b\sqrt{3} \,; a, b \in I\}$ with ordinary addition.

    f.   The set M of all 2×2 matrices with rational numbers for elements and the operation of matrix multiplication.

    5.9. ISOMORPHISM. One rudimentary algebra consists of the set $S = \{a, b, c, \ldots\}$, which is either finite or infinite, and the operation o. A second rudimentary algebra consists of the set $S' = \{a', b', c', \ldots\}$ and the operation #. We assume that the set S is closed with respect to o and the set S' is closed with respect to #. We also assume that $S \sim S'$ and that a $1 : 1$ correspondence can be established so that pairs of corresponding elements are $(a, a'), (b, b'),$ $(c, c'), \ldots$. If it happens that for every $a, b \in S$, $a \circ b = c$, and the corresponding elements $a' \# b' = c'$, then we say that there is an *isomosphism* between the two algebras, or that the two algebras are *isomorphic*.

    For example, consider the two rudimentary algebras

$$S_1 = \{0, 1, 2, 3\} \text{ with the operation } + \text{ (mod 4)};$$
$$S_2 = \{+1, i, -1, -i\} \text{ with ordinary multiplication.}$$

Since each set has four elements, $S_1 \sim S_2$, we may have the following pairs of corresponding elements:

$$(0, +1), \quad (1, i), \quad (2, -1), \quad (3, -i).$$

The multiplication tables for the two rudimentary algebras are

+ (mod 4)	0	1	2	3
0	0	1	2	3
1	1	2	3	0
2	2	3	0	1
3	3	0	1	2

$\times$	$+1$	$+i$	-1	$-i$
$+1$	$+1$	$+i$	-1	$-i$
$+i$	$+i$	-1	$-i$	$+1$
-1	-1	$-i$	$+1$	$+i$
$-i$	$-i$	$+1$	$+i$	-1

If one multiplication table were to be placed on top of the other, then corresponding elements would be coincident, not only in the top row and the first column, but also in every location within the table. Consequently, the two algebras are isomorphic.

If two systems are isomorphic they are "abstractly identical" since they are identical except for the names, or symbols, for the elements and for the description of the operation. The knowledge that two systems are isomorphic can be very useful in the mathematical analysis of an applied problem. In fact, solving an applied problem requires the construction of a mathematical system that is isomorphic to the applied problem.

5.10. FIELD. For a complete study of the structure of algebra, the reader should take a course in modern algebra. We shall give the definition of field and leave to later courses the detailed developments of the structure of algebra. The definition of field, as we shall give it, is built upon the definition of group.

Given a set of numbers

$$F = \{a, b, c, \ldots\},$$

either a finite or an infinite set, in which there are two binary operations: an "additive" operation designated by # and a "multiplicative" operation designated by &. The following axioms are to be satisfied:

5.10.1. *Additive Abelian Group.* The set F with the operation $\#$ forms an abelian group in which the identity element is denoted by z, called "zero," and the additive inverse for $a \in F$ is denoted by $\bar{a} \in F$.

5.10.2. *Multiplicative Abelian Group.* The set F with the operation $\&$ almost forms an abelian group, with the exception that the element z has no multiplicative inverse, in which the identity element is "1," called "unity" or "one," and the inverse of $a \in F$, $a \neq z$, with respect to the operation $\&$ is denoted by a'.

5.10.3. *The Distributive Law.* One form of the distributive law is satisfied:

$$a \mathrel{\&} (b \mathbin{\#} c) = (a \mathrel{\&} b) \mathbin{\#} (a \mathrel{\&} c).$$

If F is a finite set, then we have a finite field. If F is an infinite set then we have an infinite field. If F is finite with n elements then the field is finite of *order n*.

An example of a finite field of order 5 consists of the set $\{0, 1, 2, 3, 4\}$ and the operations of $+$ (mod 5) and $\times$ (mod 5).

We state and prove the following theorems. The proof, in each theorem, depends only upon the definition of a field.

5.10.4. *Theorem* 1.   The zero, z, is unique.

Proof: Assume that there are two zeros, $z \in F$ and $z^* \in F$, such that both

$$a \mathbin{\#} z = a \qquad \text{and} \qquad a \mathbin{\#} z^* = a,$$

for every $a \in F$. Consequently, each of these statements must be true both for $a = z$ and for $a = z^*$, so that the equations

$$z^* \mathbin{\#} z = z^* \qquad \text{and} \qquad z \mathbin{\#} z^* = z,$$

are both true. Since the field contains an abelian group with respect to the operation $\#$, it follows that

$$z^* \mathbin{\#} z = z \mathbin{\#} z^* = z^* = z.$$

5.10.5. *Theorem* 2.   The cancellation laws hold for each operation.

a.   The additive cancellation laws:

    i. if $a \# b = a \# c$, then $b = c$,

    ii. if $b \# a = c \# a$, then $b = c$.

b. The multiplicative cancellation laws:

    iii. if $a \,\&\, b = a \,\&\, c$, then $b = c$,

    iv. if $b \,\&\, a = c \,\&\, a$, then $b = c$.

Proof of i: We prove only the first of the additive cancellation laws, and leave the proofs of the others to the reader. The pattern of proof suggested in this part is readily adapted to the other parts. We assume that if two numbers are equal, then either number may be used in the place of the other. In particular, we shall replace the number $(a \# b)$ by $(a \# c)$ in the expression $\bar{a} \# (a \# b)$, as is indicated in the following:

$$\bar{a} \# (a \# b) = \bar{a} \# (a \# c).$$

The associative law allows us to change this equation into

$$(\bar{a} \# a) \# b = (\bar{a} \# a) \# c,$$

which becomes

$$z \# b = z \# c,$$

and reduces, finally to

$$b = c.$$

5.10.6. *Theorem* 3. $\forall\, a \in F, a \,\&\, z = z$.

Proof: We start with the expression, $a \# (a \,\&\, z)$; then

$$a \# (a \,\&\, z) = (a \,\&\, 1) \# (a \,\&\, z) = a \,\&\, (1 \# z) = a \,\&\, 1 = a.$$

The first and last member of this sequence of equalities give us,

$$a \# (a \,\&\, z) = a = a \# z.$$

When the cancellation law i is applied to the first and third members of this equation we obtain

$$a \,\&\, z = z.$$

5.10.7. *Theorem* 4. $(a \,\&\, \bar{b}) = (\bar{a} \,\&\, b) = (\overline{a \,\&\, b})$.

Proof: The definition of additive inverse provides that

$$(a \,\&\, b) \# (\overline{a \,\&\, b}) = z,$$

and from the distributive law we obtain

$$(a \mathbin{\&} b) \mathbin{\#} (a \mathbin{\&} \overline{b}) = a \mathbin{\&} (b \mathbin{\#} \overline{b}) = a \mathbin{\&} z = z.$$

Since the first members of each of these two equations are equal to z, it follows that

$$(a \mathbin{\&} b) \mathbin{\#} (\overline{a \mathbin{\&} b}) = (a \mathbin{\&} b) \mathbin{\#} (a \mathbin{\&} \overline{b}),$$

from which we obtain, by the cancellation law i,

$$(\overline{a \mathbin{\&} b}) = (a \mathbin{\&} \overline{b}).$$

Again, from the distributive law, we obtain

$$(a \mathbin{\&} b) \mathbin{\#} (\overline{a} \mathbin{\&} b) = (a \mathbin{\#} \overline{a}) \mathbin{\&} b = z \mathbin{\&} b = b \mathbin{\&} z = z,$$

so that we have

$$(a \mathbin{\&} b) \mathbin{\#} (\overline{a \mathbin{\&} b}) = (a \mathbin{\&} b) \mathbin{\#} (\overline{a} \mathbin{\&} b).$$

The cancellation law i now gives us

$$(\overline{a \mathbin{\&} b}) = (\overline{a} \mathbin{\&} b).$$

Consequently, we have established that

$$(\overline{a \mathbin{\&} b}) = (a \mathbin{\&} \overline{b}) = (\overline{a} \mathbin{\&} b).$$

5.10.8. *Theorem 5.* $(\overline{a} \mathbin{\&} \overline{b}) = (a \mathbin{\&} b).$

Proof: From Theorem 4, it follows that

$$(a \mathbin{\&} b) \mathbin{\#} (\overline{a} \mathbin{\&} b) = z.$$

From the distributive law, we obtain

$$(\overline{a} \mathbin{\&} \overline{b}) \mathbin{\#} (\overline{a} \mathbin{\&} b) = \overline{a} \mathbin{\&} (\overline{b} \mathbin{\#} b) = \overline{a} \mathbin{\&} z = z.$$

Consequently,

$$(a \mathbin{\&} b) \mathbin{\#} (\overline{a} \mathbin{\&} b) = (\overline{a} \mathbin{\&} \overline{b}) \mathbin{\#} (\overline{a} \mathbin{\&} b),$$

from which we obtain, by means of the cancellation law i,

$$(a \mathbin{\&} b) = (\overline{a} \mathbin{\&} \overline{b}).$$

PROBLEM SET 5.2

1. Prove that the set of all rational numbers with ordinary addition and ordinary multiplication comprises a field.

2. Prove that the set, K, of all real numbers with ordinary addition and multiplication comprises a field. Show that the field with rational numbers discussed in Problem 1 is a subfield of the field of real numbers.

3. Determine whether the set $\{0, 1, 2, 3, 4, 5\}$ with the operations of $+$ (mod 6) and $\times$ (mod 6) comprises a field.

4. Test the following to see if there is a field: the set $\{0, 1\}$ comprises a binary boolean algebra with the operations of $\#$ and $\&$ defined in the following multiplication tables:

#	0	1
0	0	1
1	1	1

&	0	1
0	0	0
1	0	1

5. Does the following form a field?

the set $\{a + b\sqrt{2} \; ; a, b \in R\}$,
with ordinary addition and multiplication.

6. Test to see if we have a field:

the set of numbers $\{0, 1, 2, 3, 4, 5, 6\}$
with the operations of $+$ (mod 7) and $\times$ (mod 7).

Can you predict which modular algebras will comprise fields? Is there any significance in whether the modulus is a prime or a composite number?

CHAPTER SIX

Geometries, Pure and Simple

6.1. MATHEMATICAL SYSTEMS, GEOMETRY. A *mathematical system* consists of a set S of elements which may or may not be clearly defined and certain theorems for which we seek to establish truth values. The truth value of a theorem is established by the use of logical procedures from other theorems for which truth values have been established previously. Thus, the proof of a theorem must depend on a chain of previously proved theorems. Consequently, either this chain of theorems must have no beginning, or there must be a set of theorems which are to be accepted as true without proof. Such a theorem which is accepted as true without a proof is called an *axiom*. The basis of a mathematical system, then, is a set of elements and a set of axioms. The mathematical system consists of this basis together with one or more theorems whose truth values can be established by logical procedures within the system.

Certain mathematical systems have acquired special names. For example, groups and fields, which were discussed in Chapter Five, are mathematical systems.

The mathematical systems that are to be discussed in this chapter are called *geometries*. We shall not attempt to define a geometry, since it seems more appropriate to leave this definition for more sophisticated and complete discussions of the subject. The reader will recognize a flavor of what is usually accepted as geometry in some of the systems, especially in those systems where the elements are called by such names as point, line, plane, 3-space. We shall not defend the designation of the mathematical systems discussed in this chapter as being geometries.

The elements in the basis of a geometry may be contained in a finite set or in an infinite set. If the set of elements is a finite set, then the geometry is called a *finite geometry*, or sometimes, a *miniature geometry*. If the set of elements is an infinite set, then the geometry of which it forms a base is called an *infinite geometry*. If the geometry does not include any algebraic operation, or is not referred to a coordinate system, it is called a *pure geometry*, or a *synthetic geometry*. If the geometry is referred to a coordinate system, like a cartesian plane, or a cartesian 3-space, together with algebraic operations among its elements, then it is called an *analytic geometry*. We shall examine a few examples of pure geometries in this chapter, and some examples of analytic geometries in Chapters Seven and Eight.

Whenever it is possible to do so, it is customary to construct a model for each geometry. The model may consist of a sketch based on our usual images of points, lines, planes, 3-spaces, and so on. For some geometries, the models may be constructed of subsets of natural numbers, or of letters of the alphabet, or of wires, string, sticks, and other solid objects. It must be emphasized that the construction of a model is not an essential part of the geometry, and it may not even be desirable or possible to construct a satisfactory model. The model is never, in any way, part of the proof of any theorem.

6.2. A SIMPLE FINITE PURE GEOMETRY. The first example of a finite pure geometry that we describe is the following:

Given a set S of undefined elements called *bees*. Certain subsets of S are called *hives*, so that each hive contains bees. The following axioms characterize the bees and the hives:

6.2.1 *Axiom 1.*   Every hive is a collection of bees.

6.2.2. *Axiom 2.*   Any two distinct hives have one and only one bee in common.

6.2.3. *Axiom 3.*   Every bee belongs to two and only two hives.

6.2.4. *Axiom 4.*   There are exactly four hives.

We propose three theorems, and prove that each one is true.

6.2.5. *Theorem 1.*   There are exactly six bees.

Proof: Since there are, by Axiom 4, exactly four hives we call them by the names a, b, c, d. Since, by Axiom 3, every bee belongs to two and only two hives, a nonordered pair of two distinct hives is a bee. Consequently, the only bees there can be in this system are

ab,   ac,   ad,   bc,   bd,   cd,

since these six combinations comprise all the possible nonordered pairs of hives *a*, *b*, *c*, *d*.

6.2.6. *Theorem* 2.   There are exactly three bees in each hive.

Proof: If we take hive *a* as a representative hive, the bees in hive *a* are

ab,   ac,   ad,

and no others. The same story can, obviously, be told for each of hives *b*, *c*, and *d*.

6.2.7. *Theorem* 3.   For each bee there exists exactly one other bee which is not in the same hive with it.

Proof.   We select *ab* as a representative bee. By inspection of the other bees, we find that *ab* shares a hive with each of *ac*, *ad*, *bc*, *bd*, but it does not share a hive with the bee *cd*.

We shall call two bees which do not share a hive *conjugate* bees. Thus, *ab* and *cd* are conjugate bees.

We find that *ac* and *bd* are conjugate bees, and that *ad* and *bc* are conjugate bees.

It is easy to construct a "model" for this geometry by using lines and points. One model represents the four hives by lines *a*, *b*, *c*, *d*, and the bees by the six points in which these four lines intersect, two at a time. This model is sketched below.

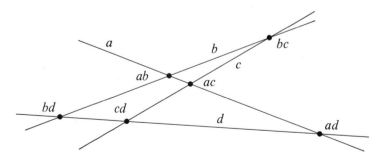

A model for the hives and bees geometry

6.3. PROPERTIES FOR A SET OF AXIOMS.   Since a geometry is a logical system the axioms defining it must be formulated with care. Each set of axioms must be consistent according to the following definition. The other two properties defined below are not essential.

6.3.1. *Consistency*.   A set of axioms is *consistent* if it does not allow for any contradiction; no axiom in the set may contradict any other axiom. This is an essential property. Every set of axioms must be consistent.

6.3.2. *Independence*.   If no axiom can be proved as a theorem by assuming the remaining axioms, then the set of axioms is *independent*. This property is not essential, but it is desirable. A set of axioms which is not independent lacks mathematical elegance.

6.3.3. *Categoricalness*.   If every model, or representation, of a mathematical system defined by a given set of axioms is essentially the same as every other possible model, or representation, of that system, then the set of axioms defining the system is said to be *categorical*. In other words, if a system is defined by a categorical set of axioms, then every pair of models of the system is isomorphic to each other. This property is not essential, and, in fact, it may not even be desirable.

6.4. A SECOND MINIATURE PURE GEOMETRY.   Finite geometries are also called *miniature* geometries. A miniature geometry with a set containing nine elements is described as follows:

The set for this miniature geometry is

$$S = \{A, B, C, D, E, F, G, H, I\},$$

where the elements are undefined. Certain subsets of S are called *m-classes*. There are eight axioms for this system:

6.4.1. *Axiom 1*.   If $A, B \in S$, then there exists an *m*-class containing both A and B.

6.4.2. *Axiom 2*.   If $A, B \in S$, and $A \neq B$, then there exists not more than one *m*-class containing both A and B.

6.4.3. *Axiom 3.* Given an *m*-class *a* and an element $P \in S, P \notin a$, there exists one *m*-class containing *P* and not intersecting *a*.

6.4.4. *Axiom 4.* Given an *m*-class *a* and an element $P \in S, P \notin a$, there exists not more than one *m*-class containing *P* and not intersecting *a*.

6.4.5. *Axiom 5.* Every *m*-class contains at least three elements of *S*.

6.4.6. *Axiom 6.* Not all of the elements of *S* belong to the same *m*-class.

6.4.7. *Axiom 7.* There exists at least one *m*-class.

6.4.8. *Axiom 8.* No *m*-class contains more than three elements of *S*.

In this geometry we construct a model before proposing any theorems. The following model may be constructed by using nine elements distributed into twelve *m*-classes:

$$(ABC) \quad (BDF) \quad (CDI) \quad (DGH)$$
$$(ADE) \quad (BEH) \quad (CEG) \quad (EFI)$$
$$(AFG) \quad (BGI) \quad (CFH)$$
$$(AHI)$$

In this model, each element belongs to four *m*-classes. Each element behaves in the same manner as every other element, so it is possible to select a single element to represent every element. In the same way, one *m*-class may be selected to represent all of the *m*-classes.

In order to see that Axiom 1 and Axiom 2 are satisfied by this model we need only to inspect the *m*-classes to ensure that each element is in one and only one *m*-class with every one of the other elements. For Axiom 3 and Axiom 4, we select the *m*-class *(ABC)* to represent all *m*-classes. The element *D* does not belong to *(ABC)*. But *D* does belong to the *m*-class *(DGH)*, which does not intersect *(ABC)*. The other *m*-classes containing *D* are *(ADE)*, *(BDF)*, and *(CDI)*, each of which intersects *(ABC)*. The remaining axioms are checked by simply inspecting the model.

6.4.9. *Theorem* 1.   The eight axioms are satisfied if and only if S contains exactly nine elements.

Proof: That the axioms are satisfied if S contains nine elements has just been demonstrated. In order to see if the axioms may be satisfied by eight elements, we try deleting one element, say A, from S. This will leave each of the m-classes in the first column of the model with two elements each, violating Axiom 5. The remaining eight m-classes do not satisfy Axioms 1 and 2. There is no way to construct a model with only eight elements. Moreover, if we add a tenth element, J, to S we find that there is no place to put it into an m-class.

We may call two m-classes that do not intersect, *conjugate m-classes*.

6.4.10. *Theorem* 2.   For each m-class there are two conjugate m-classes.

Proof: Upon inspection of the model, we find the following sets of nonintersecting conjugates:

$$\{(ABC), (DGH), (EFI) \},$$
$$\{(ADE), (BGI), \ (CFH)\},$$
$$\{(AFG), (BEH), (CDI)\},$$
$$\{(AHI), (BDF), (CEG)\}.$$

PROBLEM SET 6.1

1. Experiment with constructing other models for the geometries in Sections 6.2 and 6.4.

2. Investigate the sets of axioms in Sections 6.2 and 6.4 concerning the properties of consistency, independence, and categoricalness.

3. A miniature pure geometry is defined by a set, $S = \{A, B, C, \ldots\}$ of undefined elements, certain subsets of which are called "m-classes," and the following axioms:
    Axiom 1.   If A and B are distinct elements of S, then there exists one and only one m-class containing both A and B.

Axiom 2. Two *m*-classes having no element in common are called "conjugate" *m*-classes, and for every *m*-class there is one and only one conjugate *m*-class.

Axiom 3. There exists at least one *m*-class.

Axiom 4. Every *m*-class contains at least one element of *S*.

Axiom 5. Every *m*-class contains only a finite set of elements of *S*.

Prove that each of the following theorems is true. Construct at least one model of this geometry.

Theorem 1. Every *m*-class contains at least two elements.

Theorem 2. The set *S* contains at least four elements.

Theorem 3. The set *S* contains at least six *m*-classes.

Theorem 4. No *m*-class contains more than two elements.

6.5. A MINIATURE PROJECTIVE GEOMETRY. We shall not define the word "projective" at this time, but shall consider it just an undefined name. We have a set

$$S = \{A, B, C, \ldots\},$$

of undefined elements called *points*. Certain undefined subsets of *S* are called *lines* and each line is a collection of points. Lines will be designated by lower case letters from the beginning of the alphabet. Thus, the symbols *a, b, c, . . .* will designate lines. Certain other subsets of *S* will be called *planes*, and each plane will be a collection of lines and points. Planes will be designated by lower case letters from the second half of the alphabet. Thus, the symbols *p, q, r, . . .* will designate planes. Certain subsets of *S* which contain points, lines, and planes will be called *3-spaces*. If a point *A* belongs to line *a*, we say that *A* is on line *a*, or that *a* contains *A*, or that *a* passes through *A*. If two points *A* and *B* belong to the same line we say that *A* and *B* are *collinear*. If a point *A* belongs to two lines, we say that the lines are *concurrent* at *A*, or that the lines *intersect* at *A*.

We have the following seven axioms:

6.5.1. *Axiom 1.* There exist at least two distinct points.

6.5.2. *Axiom 2.* Two distinct points, *A* and *B*, determine one and only one line containing both *A* and *B*.

6.5.3. *Axiom 3.* If *A* and *B* are distinct points, there is at least one point distinct from *A* and *B* on the line containing *A* and *B*.

6.5.4. *Axiom 4.* If *A* and *B* are distinct points there exists at least one point not on line *AB*.

6.5.5. *Axiom 5.* If *A*, *B*, *C* are three noncollinear points and *D* is a point on *BC* distinct from *B* and *C*, and *E* is a point on *CA* distinct from *C* and *A*, then there is a point *F* on *AB* distinct from *A* and *B* such that *D*, *E*, *F* are collinear.

6.5.6. *Axiom 6.* If *A*, *B*, *C*, are three noncollinear points then there exists at least one point *D* not on the plane *ABC*.

6.5.7. *Axiom 7.* Any two distinct planes have a line in common.

Some of these axioms may at first appear to be valid in the ordinary euclidean geometry. However, Axiom 5 is not true in euclidean geometry, since it requires that every two distinct lines in a plane intersect in a point. A model for Axiom 5 made of points and lines in the usual manner is:

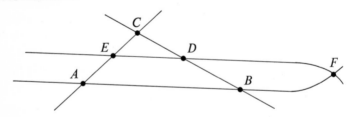

Axiom 6 requires that we define a plane. If *A*, *B*, *C* are three noncollinear points, the *plane ABC* is the set of points lying on lines joining *C* to the points of line *AB*. Since the roles of the points *A*, *B*, *C* in this definition may be interchanged it follows that three noncollinear points determine a unique plane. If *A* and *B* are distinct points of a plane *p*, then Axiom 5 requires that every point of line *AB* belong to *p*. If a plane contains a certain line, we say that the "line lies on the plane," or that the "plane passes through the line." A set of points, or a set of lines that belong to a plane, is said to be *coplanar*. As a consequence of Axiom 5, every two distinct coplanar lines intersect in a common point. Every pair of distinct concurrent lines determines a plane.

If A, B, C, D are four noncoplanar points, then the *3-space ABCD* is the set of points lying on line joining D to the points of plane ABC. Since we can interchange, or permute, the role of the points in this definition, we have that any four noncoplanar points A, B, C, D determine a unique 3-space $ABCD$.

In general, if $A_1, A_2, A_3, \ldots, A_n, A_{n+1}$ are $n + 1$ nonco$(n - 1)$-spatial points, then the *n*-space $A_1A_2A_3\cdots A_nA_{n+1}$ is the set of points lying on lines joining A_{n+1} to the points of the $(n - 1)$-space $A_1A_2A_3\cdots A_n$. Any set of $(n + 1)$ nonco$(n - 1)$-spatial points determines a unique *n*-space.

The seven axioms define a geometry which ignores both distance and parallelism. A geometry with such characteristics for points, lines, planes, and so on, is called a *projective geometry*. The set of seven axioms is not categorical, since there are representations of it which are quite different from each other. In fact it is possible to find a representation of this geometry which contains an infinite set of elements, and, as will be demonstrated in the following section, at least one representation of it which contains only a finite set of elements.

The following principles of duality are valid in this projective geometry:

The principle of duality in the plane: If, in any axiom or theorem or statement in the projective geometry of the plane, we interchange the word "point" with the word "line," then interchange the word "collinear" with the word "concurrent," and interchange "meet in" with "lie on," we shall obtain the *plane dual* of the given axiom or theorem or statement. If the original statement is true, then its plane dual is also true.

The principle of duality in 3-space: We obtain the *3-space dual* of a statement in the projective geometry of 3-space by interchanging the word "point" with the word "plane" while leaving the word "line" unchanged, and at the same time interchanging such words as "concurrent" with "coplanar," and leaving such words as "collinear" unchanged.

6.6. FANO'S FINITE PROJECTIVE GEOMETRY.   G. Fano first proposed, about 1892, a finite geometry based on the seven axioms of projective geometry, as proposed in Section 6.5. Fano's finite projective geometry contains a set of fifteen points. The fifteen points lie on thirty-five lines each of which contains three points. The thirty-five lines lie on fifteen planes, each of which contains seven points and seven lines.

Fano made use of six symbols a, b, c, d, e, f. Each point in the geometry is designated by a nonordered pair of elements from this set of six symbols, so that, for example, $ab = ba$ is one point. Hence, the fifteen points in this geometry are

$$
\begin{array}{lllll}
ab & bc & cd & de & ef \\
ac & bd & ce & df & \\
ad & be & cf & & \\
ae & bf & & & \\
af & & & &
\end{array}
$$

The thirty-five lines, each of which contains three points, are classified into two types:

Lines of type I: Each line of type I contains three points of the form (ab, bc, ca). There are twenty lines of Type I.

Lines of type II: Each line of type II contains three points of the form (ab, cd, ef). There are fifteen lines of Type II.

Any set of three points which does not belong either to type I or to type II is noncollinear. Any set of three noncollinear points determines a plane that will contain four other points.

We may construct models of the planes in Fano's finite projective geometry by "drawing" lines and points in somewhat the usual manner. However, we must remember that each line contains only three points. Perhaps, we can imagine each line to be like a wire and the points on it to be like beads strung on the wire.

The following is a model of one of the fifteen planes.

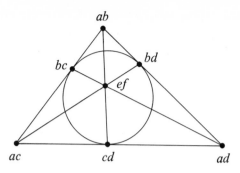

6.7. DESARGUES' THEOREM. An interesting theorem that is true both in infinite projective geometries and in Fano's finite projective geometry bears the name of the French mathematician Gerard Desargues (1593–1662). This famous theorem is stated as follows:

Desargues' theorem. If two triangles ABC and $A'B'C'$ are located either in the same plane or in different planes and are so situated that BC and $B'C'$ meet in a point L, CA and $C'A'$ meet in a point M, and AB and $A'B'$ meet in a point N, where L, M, N are collinear on line a, then the lines AA', BB', CC' are concurrent in a point O.

A "geometric" model of points and lines is sketched below:

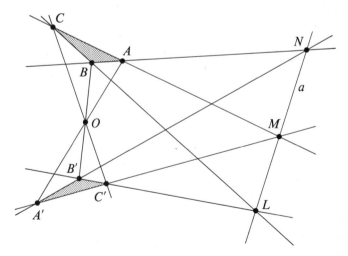

Proof: Assume that triangle ABC lies on plane p and that triangle $A'B'C'$ lies on another plane p', and that line a is the intersection of p and p'. Since AB and $A'B'$ intersect in N, the four points A, B, A', B' are coplanar. Since BC and $B'C'$ intersect in L, the four points B, C, B', C' are coplanar. Since AC and $A'C'$ intersect in M, the four points A, C, A', C' are coplanar. Since every pair of planes intersect in a line (Axiom 7), we have the following:

> plane $AA'BB'$ and plane $BB'CC'$ intersect in line BB';
> plane $BB'CC'$ and plane $AA'CC'$ intersect in line CC';
> plane $AA'BB'$ and plane $AA'CC'$ intersect in line AA'.

The three lines, AA', BB', CC' must intersect in a common point O, which is the point of intersection of the three planes $AA'BB'$, $BB'CC'$, $AA'CC'$.

The ten points and ten lines in the model for Desargues' theorem are the points and lines in which five distinct planes intersect, if three of the planes intersect in a common point O and the remaining two planes intersect in a line a.

The converse of Desargues' theorem is also true.

There are several models for Desargues' theorem in Fano's finite geometry. For an example, we may select triangles T_1 and T_2 in Fano's finite geometry, where

$$T_1 = (ab, ac, ad),$$
$$T_2 = (be, ce, de).$$

For triangles T_1 and T_2 we have the hypothesis of Desargues' theorem, since

> side (ab, ac) and side (be, ce) meet in point bc;
> side (ac, ad) and side (ce, de) meet in point cd;
> side (ad, ab) and side (de, be) meet in point bd;
> and the points (bc, cd, bd) lie on a Type I line.

Lines joining corresponding vertices of T_1 and T_2 meet in point ae, since

> ab and be are collinear with ae in a Type I line;
> ac and ce are collinear with ae in a Type I line;
> ad and de are collinear with ae in a Type I line.

PROBLEM SET 6.2

1. Construct a point and line model for each of the following Fano planes and experiment with fitting them together in a 3-space model.

   a. (ab, ac, ae);   e. (ac, ad, ae);

   b. (ab, ad, ae);   f. (ac, bd, ae);

   c. (bc, ad, ae);   g. (ab, cd, ae);

   d. (bc, bd, ae);   h. (ab, ac, ad).

2. Make a list of the remaining seven Fano planes.
   Experiment with a pattern for selecting three noncollinear points to determine the 15 planes without duplicating the planes.

3. Find two other examples, different from triangles T_1 and T_2, of pairs of Fano triangles which satisfy Desargues' theorem.

4. We are given the set of points $\{A, B, C, D, E, F, G\}$ and the following five axioms:

   6.2.4.1. There exists at least one line.
   6.2.4.2. There are exactly three points on each line.
   6.2.4.3. Not all points are on the same line.
   6.2.4.4. There is exactly one line on any two distinct points.
   6.2.4.5. There is at least one point on any two distinct lines.

   a. Set up the array of points in which the rows represent points and each column (of three elements) represents a line containing three points:

   $$
   \begin{array}{ccccccc}
   A & B & C & D & E & F & G \\
   B & C & D & E & F & G & A \\
   D & E & F & G & A & B & C.
   \end{array}
   $$

   Demonstrate that this set of seven points and seven lines satisfies the system with the five axioms.

   b. Leave the first two rows in the array as they are but change the order of points in the third row. Prove that of the remaining possible rearrangements for the third row only one arrangement provides a satisfactory model for this system. Show that the two models for this system are isomorphic.

5. In the finite geometry defined in Problem 4 delete the fourth axiom, 6.2.4.4., and replace it with the axiom:

    6.2.4.4'.   There are exactly four points on each line.

Leave the other four axioms the same as stated.

a. Prove that the set of seven points must now be replaced by a set of thirteen points. Set up the array of four rows of thirteen points in each line and each column contains the four points on each line.

b. Determine which of the possible arrays provides a representation of the system based upon the five axioms in this new system.

CHAPTER SEVEN

Analytic Geometry in the Cartesian Plane

7.1. ANALYTIC GEOMETRY. We return to the ordinary euclidean geometry, which is based on the axioms published by Euclid (about 330–275 B.C.), and which is the ordinary geometry that we learn, if that is the correct term, in secondary school.

We study euclidean geometry in the cartesian plane, so that the set of points, lines, circles, and so on, discussed in any example will be a relation. It will not only be possible, but in fact it is usually advisable, to sketch the graph for each relation. Each relation can be defined by a statement in the language of ordinary algebra, and it will be possible to do the standard manipulations of ordinary algebra in order to make studies of the properties of the given relation.

The study of geometric properties by means of algebraic statements and methods is called *analytic geometry*.

7.2. DISTANCE BETWEEN TWO POINTS. Given two points

$$P_1 = (x_1, y_1) \qquad \text{and} \qquad P_2 = (x_2, y_2)$$

in the cartesian plane, there is a real number, designated by $d(P_1, P_2)$ and called the *distance* between P_1 and P_2, defined with the following properties:

7.2.1. $d(P_1, P_2) \geq 0$;

7.2.2. $d(P_1, P_2) = 0 \leftrightarrow P_2$ coincides with P_1;

7.2.3. $d(P_1, P_2) = d(P_2, P_1)$;

7.2.4. $d(P_1, P_2) + d(P_2, P_3) \geq d(P_1, P_3)$;

7.2.5. $d(P_1, P_2) = \sqrt{(x_2 - x_1)^2 + (y_2 - y_1)^2}$.

If P_1 and P_2 have the same abscissa, or x coordinate, then

$$d(P_1, P_2) = \sqrt{(y_2 - y_1)^2} = |y_2 - y_1|.$$

If P_1 and P_2 have the same ordinate, or y coordinate, then

$$d(P_1, P_2) = \sqrt{(x_2 - x_1)^2} = |x_2 - x_1|.$$

7.3. THE DIRECTION OF A LINE EXPRESSED BY THE SLOPE.

If we have two points $P_1 = (x_1, y_1)$ and $P_2 = (x_2, y_2)$, then there is a line determined by P_1 and P_2. We are interested in finding an algebraic, or numerical, way to describe the direction of this line with respect to the reference frame established by the coordinate axes.

Since a line has constant direction, we may seek a means of defining a measure of its direction at any part of the line. We define the following two numbers:

$(y_2 - y_1)$ = vertical change on the line from P_1 to P_2 = *rise;*
$(x_2 - x_1)$ = horizontal change on the line from P_1 to P_2 = *run.*

The direction of the line determined by P_1 and P_2 is described by the single number obtained as the ratio of the rise to the run. This number is called the *slope* of the line, where

$$\text{slope} = \frac{(y_2 - y_1)}{(x_2 - x_1)} = \frac{\text{rise}}{\text{run}},$$

so that the slope expresses the rate of rise of the line per unit of run.

We see, from the definitions, that the rise and the run may be either positive or negative, so that the slope may be positive or negative. If the slope is positive, then the direction of the line is such that the line rises as one goes on it from left toward the right. If the slope is negative then the line slants downward as one goes on it from the left toward the right. If the slope is 0, then the rise must be 0, so that the line is horizontal, or parallel to the x axis. In a vertical line, the run is always 0, so the slope of a vertical line is not defined.

If the line passes through the point $P_1 = (x_1, y_1)$ with slope p/q, then a second point on the line is

$$(x_1 + \text{run}, y_1 + \text{rise}) = (x_1 + q, y_1 + p).$$

Moreover, since $kp/kq = p/q$, for every $k \in K$, $k \neq 0$, it follows that each of the points

$$(x_1 + kq, y_1 + kp),$$

for every real number k, will also lie on the same line.

For an example, the line which passes through the point $(3, 2)$ with slope $4/5$ passes through the points $(3 + 5k, 2 + 4k)$ for every $k \in K$.

PROBLEM SET 7.1

1. Find the rise, the run, and the slope and sketch the graph for the line AB, if
   a. $A = (1, 2)$, $B = (2, 5)$;
   b. $A = (-5, 3)$, $B = (6, 3)$;
   c. $A = (102, 3)$, $B = (-3, 105)$;
   d. $A = (-3, -6)$, $B = (-5, -7)$;
   e. $A = (7, 0)$, $B = (-3, -7)$.

2. Find the distance $d(A, B)$ in each part of Problem 1.

3. Given the four points

   $$A = (3, 5), \quad B = (8, 6), \quad C = (10, 10), \quad D = (5, 9).$$

   Describe in detail the quadrilateral $ABCD$.

3. Locate and name three other points on each line determined by the given conditions:
   a. passing through $(-3, 4)$ with slope $4/5$;
   b. passing through $(-2, -8)$ with slope $-7/9$;
   c. passing through $(6, -5)$ and $(-2, -5)$;
   d. passing through $(2, 5)$ and $(9, 11)$;
   e. passing through $(-4, 5)$ parallel to the line with slope $3/7$.

7.4. LINE SEGMENTS.   Given the two distinct points

$$P_1 = (x_1, y_1) \qquad \text{and} \qquad P_2 = (x_2, y_2)$$

the portion of the line determined by P_1 and P_2 which lies between P_1 and P_2 is called the *line segment* P_1P_2. The points P_1 and P_2 are called the *end points* of the line segment.

The slope of line P_1P_2 is

$$\frac{y_2 - y_1}{x_2 - x_1} = \frac{p}{q} = m,$$

where $p = k(y_2 - y_1)$ and $q = k(x_2 - x_1)$. We know from Section 7.3 that the point

$$P = (x_1 + q, y_1 + p) = (x_1 + k(x_2 - x_1), y_1 + k(y_2 - y_1)),$$

for every $k \neq 0$, lies on the line determined by P_1 and P_2. The three line segments determined by P_1, P_2, and P have the following lengths:

$$
\begin{aligned}
d(P_1, P_2) &= \sqrt{(x_2 - x_1)^2 + (y_2 - y_1)^2}; \\
d(P_1, P) &= \sqrt{k^2(x_2 - x_1)^2 + k^2(y_2 - y_1)^2} \\
&= |k|\, d(P_1, P_2); \\
d(P, P_2) &= \sqrt{(1 - k)^2(x_2 - x_1)^2 + (1 - k)^2(y_2 - y_1)^2} \\
&= |1 - k|\, d(P_1, P_2).
\end{aligned}
$$

If $x_1 < x_2$ and if $0 < k < 1$, then the point P lies on the line segment between P_1 and P_2 and the point P divides the line segment P_1P_2 in the ratio of $k : (1 - k)$.

In particular, if we take $k = 1/2$, then the ratio
$$k : (1 - k) = 1/2 : 1/2 = 1 : 1,$$
so that the point P is the midpoint of the line segment P_1P_2. The coordinates of the midpoint, $P_{1:1}$, are

$$
\begin{aligned}
P_{1:1} &= (x_1 + \tfrac{1}{2}(x_2 - x_1), y_1 + \tfrac{1}{2}(y_2 - y_1)) \\
&= \left(\frac{x_2 + x_1}{2}, \frac{y_2 + y_1}{2} \right).
\end{aligned}
$$

The point that divides the line segment P_1P_2 in the ratio $p : q$ is obtained by using for k the value $p/(p + q)$, so that the coordinates of the point $p : q$ are

$$P_{p:q} = \left(\frac{px_2 + qx_1}{p + q}, \frac{py_2 + qy_1}{p + q}\right).$$

For an example, if $P_1 = (-3, 5)$ and $P_2 = (7, -11)$, then the line segment P_1P_2 may be divided in certain ratios, by the following points:

$$P_{1:1} = \left(\frac{-3 + 7}{2}, \frac{5 - 11}{2}\right) = (2, -3) = \text{midpoint};$$

$$P_{3:1} = \left(\frac{21 - 3}{3 + 1}, \frac{-33 + 5}{3 + 1}\right) = (9/2, -7) = \begin{array}{l}\text{the point that is three-}\\ \text{fourths of the way from}\\ P_1 \text{ to } P_2;\end{array}$$

$$P_{1:2} = \left(\frac{7 - 6}{3}, \frac{-11 + 10}{3}\right) = (1/3, -1/3) = \begin{array}{l}\text{the point that is one-}\\ \text{third of the way from}\\ P_1 \text{ to } P_2.\end{array}$$

7.5. THE EQUATION OF A LINE. A line is determined either by two points or by one point and the direction of the line. The direction of the line is expressed by the slope.

An open statement in two variables, which are usually x and y, that is true only for the points lying on a certain line is called the *equation of the line*.

If we know two points on a line, then its slope can be calculated immediately. Consequently, for each specific line we must know its slope and at least one point belonging to the line. In order to derive the equation for a certain line we must express the open statement that will require that the point $P = (x, y)$ lie on the line. Since a line has constant slope, the slope between $P = (x, y)$ and the given point on the line must be equal to the known slope of the line. In particular, the equation of the line determined by $P_1 = (x_1, y_1)$ and $P_2 = (x_2, y_2)$ is determined as follows:

(1) the slope of the line is $\frac{y_2 - y_1}{x_2 - x_1}$;

(2) the slope between a point $P = (x, y)$ and P_1 is $\frac{y - y_1}{x - x_1}$;

(3) the open statement which requires that $P = (x, y)$ lie on the line determined by P_1 and P_2 must say that these two slopes are the same, that is

$$\frac{y - y_1}{x - x_1} = \frac{y_2 - y_1}{x_2 - x_1},$$

which is the equation of the line.

In the same manner, we determine that the line that passes through $P_1 = (x_1, y_1)$ with slope m, has the equation

$$\frac{y - y_1}{x - x_1} = m,$$

which simplifies algebraically to $y - y_1 = m(x - x_1)$.

The equation of a horizontal line passing through $P_1 = (x_1, y_1)$ is $y - y_1 = 0$, or $y = y_1$.

The equation of a vertical line through $P_1 = (x_1, y_1)$ is $x - x_1 = 0$, or $x = x_1$.

In general, an equation of the form $y = k$, a constant, is a horizontal line, and an equation of the form $x = h$, a constant, is a vertical line.

PROBLEM SET 7.2

1. Find the midpoint of the line segment AB, where
   a. $A = (3, 5)$, $B = (5, 7)$;
   b. $A = (-4, 6)$, $B = (8, -12)$;
   c. $A = (-a, b)$, $B = (a, -b)$;
   d. $A = (p, q)$, $B = (r, s)$;
   e. $A = (7, -11)$, $B = (3\sqrt{2}, \sqrt{2})$.

2. Determine the point
   a. which is 2/3 of the way from $(9, -5)$ to $(12, 8)$;
   b. which is 7/12 of the way from $(-4, -6)$ to $(-3, -1)$.

3. Find the equation for each of the lines:
   a. connecting the points $(1, 2)$ and $(3, -5)$;
   b. passing through $(3, 4)$ with slope $-2/3$;
   c. with slope $-5/6$ and y intercept $(0, 12)$;
   d. passing through $(2 - \sqrt{5}, 3 + \sqrt{5})$ with slope $\sqrt{3}$.

7.6. FAMILY OF LINES.   The equation of a line can always be manipulated algebraically into the form

$$a_1 x + a_2 y + a_3 = 0,$$

where a_1, a_2, a_3 are real number constants. We note that the equation of a line is always a first degree equation in both the variables x and y. For this reason, a first degree equation in two variables is called a *linear equation*.

For convenience, we may replace the left member by the single symbol m_1, so that

$$m_1 = a_1 x + a_2 y + a_3 = 0,$$

is the equation of the line.

If we have two lines, L_1 and L_2, we may represent them by sets of points in the cartesian plane:

$$L_1 = \{(x, y) ; a_1 x + a_2 y + a_3 = m_1 = 0\}, \quad \text{and}$$
$$L_2 = \{(x, y) ; b_1 x + b_2 y + b_3 = m_2 = 0\}.$$

Then each of the sets of points

$$L_3 = \{(x, y) ; p m_1 + q m_2 = 0, p, q \in K, p, q \text{ not both } 0\}$$

must contain the point of intersection $L_1 \cap L_2$, since any point (x, y) which satisfies simultaneously both $m_1 = 0$ and $m_2 = 0$ will also have to satisfy $p m_1 + q m_2 = 0$, for every $p, q \in K$, not both zero. Consequently, for each set of values for p, q the equation $p m_1 + q m_2 = 0$ will be the equation of a line which will pass through the point of intersection $L_1 \cap L_2$. The set of lines represented by L_3 is called the *family* of lines through the intersection of L_1 and L_2.

The concept of family of lines, or of family of other types of relation is a very useful concept. In general, a family is a set of lines, or other curves, that share at least one common family characteristic Examples of other families of lines are:

(a) $y = 2x + k$ is the family of lines with slope 2;
(b) $y = mx + 3$ is the family of lines with y intercept at $(0, 3)$;
(c) $5x - 7y = k$ is the family of lines with slope 5/7;

Given two lines, L_1 and L_2, which are not parallel. Suppose that the lines intersect in the point

$$L_1 \cap L_2 = (h, k).$$

The usual procedure in ordinary algebra is to determine, by means of some sort of algebraic manipulations, those particular members of the family $pm_1 + qm_2 = 0$ that will give the required information that (h, k) is the point of intersection. This information is readily determined by finding the particular members of the family, one of which is the vertical line $x = h$, and the other is the horizontal line $y = k$.

For an example, given the two lines

$$m_1 = 3x - y - 1 = 0,$$
$$m_2 = x + 2y - 12 = 0.$$

We find, perhaps after a few trials, that

$2m_1/7 + m_2/7 = x - 2 = 0 =$ the required vertical line;
$-m_1/7 + 3m_2/7 = y - 5 = 0 =$ the required horizontal line.

Since each of these lines belongs to the family that has the family characteristic that every member passes through the point of intersection of the two given lines $m_1 = 0$ and $m_2 = 0$, we have discovered that the point of intersection is $(2, 5)$.

A comment here: in elementary algebra, the "point" of intersection of two given lines is frequently given in the form "$x = h$ and $y = k$." This form of answer is technically incorrect. If the *point* of intersection is the required result, then it should be given as "(h, k)."

7.7. DETACHED COEFFICIENT METHODS FOR FINDING THE INTERSECTION OF A PAIR OF LINES. It is assumed that the reader has been introduced to the usual techniques for solving for the point of intersection of a pair of lines, or of solving for the point of intersection of three planes in 3-space. We shall examine a technique that is similar to the ordinary one, but employs only the detached coefficients of the terms in the equations.

We want to find the point of intersection of two lines, $L_1 \cap L_2$, where

$$L_1 = \{(x, y) \; ; \; a_1x + a_2y + a_3 = 0\},$$
$$L_2 = \{(x, y) \; ; \; b_1x + b_2y + b_3 = 0\},$$

or, in the language of ordinary algebra, we want to solve the simultaneous linear equations

$$a_1x + a_2y + a_3 = 0,$$
$$b_1x + b_2y + b_3 = 0.$$

We detach the coefficients and, without changing their positions, place them in a 2×3 matrix,

$$\begin{pmatrix} a_1 & a_2 & a_3 \\ b_1 & b_2 & b_3 \end{pmatrix}.$$

In order to find the point, (h, k), in which the two lines intersect, we must first find the horizontal and the vertical members of the family of lines through (h, k). That is, we must first find the equations

$$x - h = 0,$$
$$y - k = 0,$$

for which the matrix of the detached coefficients is

$$\begin{pmatrix} 1 & 0 & -h \\ 0 & 1 & -k \end{pmatrix}.$$

Our task, then, is to transform the matrix of detached coefficients of the original equations to the latter matrix. This may be done by means of *elementary row operations* applied to the first matrix.

The elementary row operations for matrices are:

7.7.1.   We may multiply each element of a row by the same non-zero constant.

7.7.2.   We may add to the elements of a certain row a constant multiple of the corresponding elements of another row.

The new matrix that results from applying a row operation is not equal to the first matrix. The result of applying a row operation is a transformation of the first matrix.

For an example, we solve for the intersection of the two lines

$$3x - y = 1,$$
$$x + 2y = 12.$$

The matrix of detached coefficients is—note that we may leave the constant term on either side so long as we remember where it is—

$$\begin{pmatrix} 3 & -1 & 1 \\ 1 & 2 & 12 \end{pmatrix}.$$

The first row operation we apply is a combination of both operations: multiply the elements in the first row by 2; then add the elements of the second row to the result, to obtain the transformed matrix:

$$\begin{pmatrix} 7 & 0 & 14 \\ 1 & 2 & 12 \end{pmatrix}.$$

Next, in the transformed matrix, multiply the elements in the first row by 1/7, then add -1 times the resulting elements to the corresponding elements in the second row, to produce the matrix

$$\begin{pmatrix} 1 & 0 & 2 \\ 0 & 2 & 10 \end{pmatrix}.$$

Finally, multiply the elements in the second row by 1/2 to obtain

$$\begin{pmatrix} 1 & 0 & 2 \\ 0 & 1 & 5 \end{pmatrix}.$$

Since the final matrix is that for the detached coefficients of

$$x = 2,$$
$$y = 5,$$

the point of intersection of the given lines is $(2, 5)$.

This procedure is easily extended to solving sets of linear equations in n variables, where n is any natural number.

7.8. PARALLEL LINES, PERPENDICULAR LINES. The line passing through the point $P_1 = (x_1, y_1)$ with slope p/q has the equation

$$y - y_1 = \frac{p}{q}(x - x_1),$$

which can be manipulated algebraically to the simplified form

$$px - qy = k,$$

where $k = px_1 - qy_1$ is a constant.

We notice that in the left member of the simplified form of the equation of this line, the slope p/q appears in the coefficients of x and y. In fact, we see that the slope is actually the coefficient of x divided by the negative of the coefficient of y. For example, the slope of the line $2x - 3y = 7$ is $2/3$.

Since two parallel lines have the same direction, and consequently they have the same slope, then the coefficients of x and y in the equations of two parallel lines must be the same, or they may be constant multiples. Thus, the line $ax + by = c$ is parallel to the line $ax + by = d$, and to the line $kax + kby = f$.

For example,

$3x - 2y = 5$ and $6x - 4y = 9$ are parallel with slope $3/2$;
$y = 3x - 6$ and $3x - y = 10$ are parallel with slope 3.

A second point on the line passing through $P_1 = (x_1, y_1)$ with slope p/q is the point

$$A = (x_1 + q, y_1 + p).$$

If we construct a line through P_1 perpendicular to the given line and designate the slope of this perpendicular line by a/b, then the point

$$B = (x_1 + b, y_1 + a)$$

lies on this perpendicular line. The three points P_1, A, B are the vertices of a right triangle with its right angle at P_1. The well-known pythagorean theorem, named after Pythagoras, a half mythical figure of approximately 500 B.C., gives us that

$$d(P_1, A)^2 + d(P_1, B)^2 = d(A, B)^2, \quad \text{or}$$
$$(q^2 + p^2) + (b^2 + a^2) = (b - q)^2 + (a - p)^2.$$

The latter equation reduces algebraically to

$$-2bq - 2ap = 0,$$

from which we obtain the two important results

$$a/b = -q/p \quad \text{and} \quad (a/b)(p/q) = -1.$$

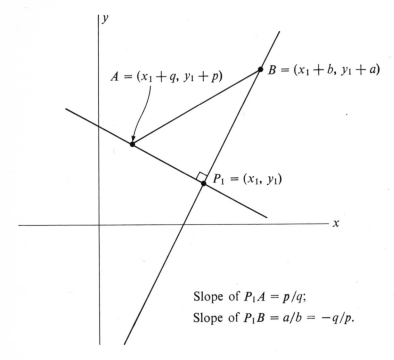

Slope of $P_1A = p/q$;
Slope of $P_1B = a/b = -q/p$.

Slopes of perpendicular lines

From this we obtain the information that if two lines are perpendicular, then the product of their slopes is -1, or, in other words, if line b is perpendicular to line a then the slope of line b is the negative reciprocal of the slope of line a.

If the slope of line a is p/q, then its equation has the form

$$px - qy = k.$$

If line b is perpendicular to line a, then the slope of b is $-q/p$, and its equation must have the form

$$qx + py = t.$$

Consequently, we need only to interchange the coefficients of x and y

and change the sign of one coefficient in the equation of a given line in order to obtain the equation of a line perpendicular to it.

For example, the lines $3x - 2y = 5$ and $2x + 3y = 7$ are perpendicular.

PROBLEM SET 7.3

1. Find the slope, a point on the line, and sketch the graph of each line:
   a. $y = 3x - 8$;
   b. $5x + 7y = 4$;
   c. $x/5 - y/9 = 1$;
   d. $3x - 11y = 60$.

2. Use the matrix of detached coefficients and elementary row operations to find the point of intersection for each of the following pairs of lines:

   a. $3x - 4y = -25$,
      $5x + 7y = 4$;
   b. $5x + 2y = 1$,
      $2x - 5y = 17$;
   c. $3x + 4 = 4y$,
      $2y + 9x = 9$;
   d. $3x - 2y = 12$,
      $5x + 2y = 4$;
   e. $x + 3y = 10$,
      $3x + y = 6$;
   f. $7 = 9y + 6x$,
      $14 - 6y = -3x$.

3. Find the equation of the line
   a. passing through $(2, -3)$ parallel to $5x + 7y = 8$;
   b. passing through $(-5, 7)$ perpendicular to $3x - 5y = 9$;
   c. passing through the intersection of $2x + 3y = 5$ and $7x - 3y = 4$, perpendicular to $5x + 2y = 1$.

7.9. THE PROOF OF SELECTED PLANE GEOMETRY THEOREMS BY ANALYTIC GEOMETRY METHODS. Certain selected theorems in plane euclidean geometry can be proved very easily by using analytic geometry. The first step in such a proof is to make a sketch of the lines and points mentioned in the hypothesis of the theorem, and then to place the reference frame of coordinate axes in the most convenient location. The best location for the axes is usually that which makes the coordinates of the points the simplest, but it may depend upon the conclusion of the theorem.

As a general rule, if the theorem involves a scalene triangle it is best to place the axes so that the vertices of the triangle are at the points $(0, 0)$, $(a, 0)$, (b, c). If the theorem mentions the midpoints of the sides of the triangle, then it will be better to place the axes and choose the unit distance so that the vertices are at the points $(0, 0)$, $(2a, 0)$, $(2b, 2c)$. If the theorem is concerned with a parallelogram, then the coordinates of the vertices should be such that opposite sides are parallel. If the theorem contains a circle, then it would probably be best to place the origin of the reference frame at the center of the circle.

We shall state and prove by analytic methods three of the well-known theorems from euclidean plane geometry.

7.9.1. *Theorem* 1.   The line joining the midpoints of two sides of a triangle is parallel to the third side and its length is one half the length of the third side.

Proof by analytic geometry: We place the axes so that the vertices of the given triangle are at

$$A = (0, 0), \quad B = (2a, 0), \quad C = (2b, 2c).$$

The graph of triangle ABC is:

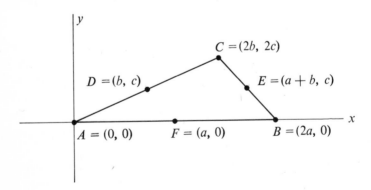

The midpoints of the sides are $D = (b, c)$, $E = (a + b, c)$, $F = (a, 0)$. The line segment DE joins the midpoints of sides AC and BC. The slope of DE is 0 and the slope of Ab is 0, so DE is parallel to AB. The length of AB is $2a$, and the length of DE is a.

7.9.2. *Theorem* 2.   The medians of a triangle meet in a point which is 2/3 of the way from each vertex to the midpoint of the opposite side.

Proof by analytic geometry: We shall use the same graph as we used in the proof of theorem 1. The medians are the line segments AE, BD, and CF. We find the equation of each line which contains a median:

AE: $y/x = c/(a + b)$,   or $cx - (a + b)y = 0$;
BD: $y/(x - 2a) = -c/(2a - b)$,   or $cx + (2a - b)y = 2ac$;
FC: $y/(x - a) = 2c/(2b - a)$,   or $2cx - (2b - a)y = 2ac$.

When we solve the first two equations together we find that the point of intersection is $(2(a + b)/3, 2c/3)$, and this point lies on the line containing median FC since its coordinates satisfy the equation of the line containing FC. If we designate this point of intersection by M, then, by methods discovered in Section 7.4, we obtain that

$$d(A, M) = 2\, d(M, E) = \tfrac{2}{3}\, d(A, E);$$
$$d(B, M) = 2\, d(M, D) = \tfrac{2}{3}\, d(B, D);$$
$$d(C, M) = 2\, d(M, F) = \tfrac{2}{3}\, d(C, F).$$

7.9.3. *Theorem* 3.   The opposite sides in a parallelogram are equal in length.

Proof by analytic geometry: A parallelogram is defined as a quadrilateral in which opposite sides are parallel. So, we place the reference frame so that one vertex is at the origin and one side lies on the x axis. Thus, two vertices are: $A = (0, 0)$ and $B = (a, 0)$. The third vertex will be at $D = (b, c)$. The fourth vertex, C, must be at the point where side AB will be parallel to side DC, and side BC will be parallel to side AD. We may temporarily let $C = (x, y)$. Then, if DC is to be parallel to AB, the slope of DC must be 0, so $y = c$. Since the slope of AD is c/b and the slope of BC, so far, is $c/(x - a)$, we must have $c/(x - a) = c/b$, from which we calculate that $x = (a + b)$. Then, the coordinates of C must be $C = (a + b, c)$ in order that $ABCD$ will be a parallelogram. Now, the length of AD is $d(A, D) = \sqrt{b^2 + c^2}$ and the length of BC is $d(B, C) = \sqrt{b^2 + c^2}$, so AD and BC are equal in length. The length of AB is $d(A, B) = a$, and the length of DC is $d(B, C) = a$, so AB and DC are equal in length.

The graph for Theorem 3 is

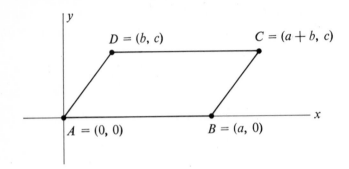

PROBLEM SET 7.4

Prove each of the following theorems by using analytic geometry.

1. The altitudes of a triangle are concurrent.

2. The lines joining the midpoints of an isosceles trapezoid, taken in order, form either a rhombus or a square.

3. The midpoint of the hypotenuse of a right triangle is equidistant from all three vertices of the triangle.

4. The diagonals of a parallelogram bisect each other.

5. The perpendicular bisectors of the sides of a triangle are concurrent in a point which is equidistant from the three vertices of the triangle.

6. Any point lying on the perpendicular bisector of a line segment is equidistant from the ends of the line segment.

7. The lines connecting, in order, the midpoints of the sides of any convex quadrilateral form a parallelogram.

8. Find other theorems from plane synthetic euclidean geometry which are easily proved by analytic geometry methods. Note that

there are many standard theorems which are more easily proved by analytic geometry, while other theorems are proved more easily by the usual synthetic methods.

7.10. CONE, CONIC SECTIONS.   A circle, C, with radius r and center at the point O lies on a horizontal plane. A vertical line OV is drawn perpendicular to the plane at the center O of the circle. We select a point V on this vertical line and construct the lines from V to points on the circumference of C. The set of all these lines forms a surface in 3-space which is called a *right circular cone*, or, more briefly, a *cone*. The line OV is called the *axis* of the cone and V is called the *vertex* of the cone. Each one of the lines through V and a point on C is called an *element* of the cone. Each element lies entirely on the surface of the cone.

A sphere which is contained inside the cone so that it is tangent to the cone at the points of a circle is called a *Dandelin sphere*, after the Belgian geometer Germinal Dandelin (1794–1847).

Whenever a cone is cut, or sliced, by a plane, the slicing plane and the cone intersect in a curve the shape of which depends on the manner in which the slice is made. Such a curve is called a *conic section*. The conic sections include the *circle*, the *parabola*, the *ellipse*, the *hyperbola*, and straight lines. The conic sections will be discussed in detail in the succeeding four sections.

7.11. THE CIRCLE.   When the slicing plane is perpendicular to the axis of the cone the intersection with the cone is a *circle*, unless the slicing plane passes through the vertex V. In the latter case, the intersection is the point V, which may be called a *point circle*. The two Dandelin spheres that are tangent to the interior of the cone both are tangent to circle C at its center O. If P is any point on C and VP is the element of the cone containing P, then PV is tangent to the Dandelin sphere D at the point B and PV is tangent to the Dandelin sphere E at point A. PO is tangent to each Dandelin sphere at O. It follows that $PO = PB = PA$. The Dandelin sphere with center at D is tangent to the cone at every point of the circle G, and G lies on a plane which is perpendicular to the axis of the cone. The Dandelin sphere with center

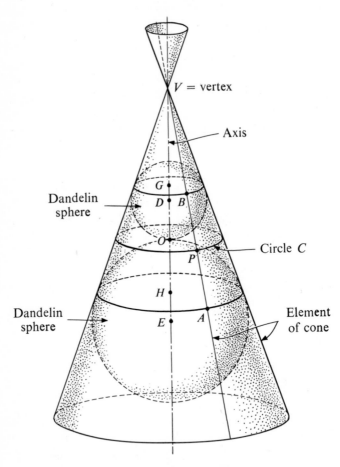

The right circular cone

at E is tangent to the cone at every point of the circle H and the plane containing H is perpendicular to the axis of the cone. Consequently, the plane containing circle G is parallel to the plane containing circle H, and each is parallel to the slicing plane containing circle C. It follows that $PA = PB$ for every point P on circle C and on the corresponding element of the cone. Since $PO = PA = PB$, it follows that every point on circle C is at a constant distance from the center O. This constant distance is called the *radius* of the circle.

If, now, we replace the slicing plane by a cartesian plane in which the unit distance and the axes are selected so that the center of C is at the origin and the radius is r distance units, we may derive the equation of the circle. The equation of the circle must express the equation which must be true for every point $P = (x, y)$ lying on the circle. The sketch of the cartesian plane containing circle C is

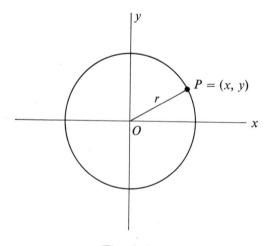

The circle

Since $d(P, O) = r$ for every point P on circle C, it follows that the equation of the circle is

$$\sqrt{x^2 + y^2} = r, \quad \text{or,}$$
$$x^2 + y^2 = r^2.$$

It is easy to see that if we should prefer to set the axes so that the

center of the circle is at the point (h, k), then the equation of the circle will be

$$(x - h)^2 + (y - k)^2 = r^2.$$

7.12. THE PARABOLA.   When the cone is sliced by a plane parallel to one and only one element, but not passing through the vertex V, the intersection with the cone is called a *parabola*. In this case there can be but one Dandelin sphere D, which is tangent to the cone in the circle with center at S and to the slicing plane at F.

We select a point P lying on the parabola and on the element VP of the cone. PF is tangent to the Dandelin sphere D at F, and PV is tangent at C to this same sphere, so

$$PF = PC.$$

The slicing plane and the plane containing circle S intersect in the line LM. The line LM is called the *directrix* of the parabola.

We construct PA perpendicular to LM at point A and construct PG perpendicular to the circle containing circle S. Then triangle PGC has a right angle at G and triangle PGA has a right angle at G. The angle PCG is the angle between the plane of circle S and an element of the cone, and angle PAG is the angle between the slicing plane and the plane of circle S. Since the slicing plane is parallel to an element of the cone,

$$\angle PCG = \angle PAG,$$

and this is valid regardless of the choice of the point P on the parabola. Since the two right triangles share the common side PG, it follows that

$$\triangle PCG \sim \triangle PAG, \quad \text{and}$$

consequently,

$$PA = PC = PF.$$

Thus, we have discovered the fundamental property of the parabola:

A parabola is the set of points that are equidistant from a fixed focus and a fixed directrix.

Now, we replace the slicing plane by a cartesian plane in which the unit distance is chosen and the axes are placed so that the focus is at the point

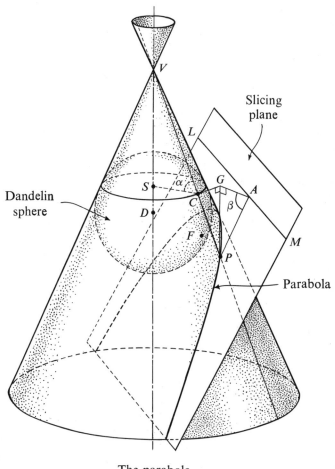

Slicing plane

Dandelin sphere

Parabola

The parabola

$$F = (p, 0), \qquad p > 0,$$

and the directrix is in the line whose equation is

$$x = -p.$$

Then, the equation of the parabola must require that the point $P = (x, y)$ be equidistant from $F = (p, 0)$ and the directrix, $x = -p$. Thus, the equation of the parabola is

$$x + p = \sqrt{(x - p)^2 + y^2},$$

which reduces by simple algebraic manipulations to

$$y^2 = 4px.$$

The line $y = 0$ is the *axis* of this parabola and the point $V = (0, 0)$ is its *vertex*.

A graph of the parabola just described is

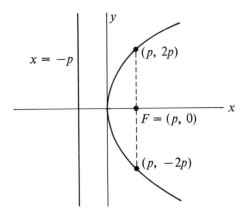

The line segment that is perpendicular to the axis at the focus with its end points in the parabola has end points at $(p, 2p)$ and $(p, -2p)$ and its length is $4p$, the coefficient of the x term in the equation. This line segment is called the *latus rectum*.

We leave to the reader the task of verifying the following facts about the parabola:

(1)   If the axis is the y axis, the focus is at $(0, p)$, and the directrix is in the line $y = -p$, then the equation of the parabola is

$$x^2 = 4py.$$

(2)   If the vertex is at (h, k), the focus is at $(h + p, k)$ and the directrix is in the line $x = h - p$, then the equation of the parabola is

$$(y - k)^2 = 4p(x - h).$$

7.13. THE ELLIPSE.   When the cone is sliced by a plane not through V but intersecting all the elements of one nappe (that is, one of the "halves") of the cone, the intersection with the cone is called an *ellipse*. In this case there are two Dandelin spheres, one of which is tangent to the cone in the circle U and to the slicing plane at the point F_1, and the other of which is tangent to the cone in the circle W and to the slicing plane at the point F_2. The plane containing circle U is parallel to the plane containing the circle V.

We select a point P on the ellipse lying on the element VP. VP is tangent to the Dandelin sphere S at A, where it touches circle U, and is tangent to the Dandelin sphere T at B where it touches circle W. The plane containing circle U is parallel to the plane containing circle W. Then,

$$PF_1 = PA, \quad \text{and}$$
$$PF_2 = PB.$$

From these equalities we obtain

$$PF_1 + PF_2 = PA + PB = AP + PB = AB.$$

The length of AB must be the same regardless of the choice of the point P on the ellipse. Thus, we have the fundamental property of the ellipse:

An ellipse is the set of all points in a plane which are so located that the sum of the distances from each point to two fixed points in the plane is constant.

The two points F_1 and F_2 are called the *foci* of the ellipse.

Line d_1 in which the slicing plane intersects the plane of circle U and line d_2 in which the slicing plane intersects the plane of circle W are called the *directrices* of the ellipse. The ratio of the distance from a point P to the focus to the distance from P to a directrix is called the *eccentricity*. We shall leave to the reader the pleasure of investigating this interesting property. The line joining F_1 and F_2 is perpendicular to each directrix.

We now replace the slicing plane by a cartesian plane in which the

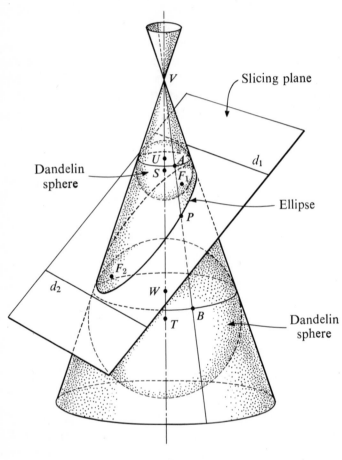

The ellipse

unit distance is selected and the axes are placed so that the foci are

$$F_1 = (c, 0) \qquad \text{and} \qquad F_2 = (-c, 0), \qquad c > 0.$$

We find that it is convenient to designate the sum of the distances from a point $P = (x, y)$ to the foci by the constant $2a$, so that

$$d(P, F_1) + d(P, F_2) = 2a$$

expresses the relation which must be satisfied for a point P to lie on the ellipse. When we translate this equation into algebraic language, we obtain

$$\sqrt{(x - c)^2 + y^2} + \sqrt{(x + c)^2 + y^2} = 2a.$$

This equation reduces by elementary algebraic manipulation to

$$\frac{x^2}{a^2} + \frac{y^2}{(a^2 - c^2)} = 1.$$

It is customary to replace $a^2 - c^2 = b^2$, and arrive at the final form for the equation of the ellipse:

$$\frac{x^2}{a^2} + \frac{y^2}{b^2} = 1.$$

The ellipse cuts the x axis at $(-a, 0)$ and $(a, 0)$ and these are the end points of the *major axis* of the ellipse. The ellipse cuts the y axis at the points $(0, -b)$ and $(0, b)$ and these are the end points of the *minor axis*. It is assumed here that $a > b$. The origin is the *center* of the ellipse.

The graph of the ellipse just discussed is

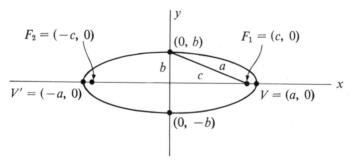

We let the reader verify the following facts about ellipses:

(1)   If the foci are on the y axis at $(0, c)$ and $(0, -c)$ then the equation of the ellipse becomes

$$\frac{x^2}{b^2} + \frac{y^2}{a^2} = 1.$$

(2)   If the center is at (h, k) and the foci are at $(h - a, k)$ and $(h + a, k)$ then the equation of the ellipse becomes

$$\frac{(x - h)^2}{a^2} + \frac{(y - k)^2}{b^2} = 1.$$

(3)   If the center of the ellipse is at (h, k) and the foci are at $(h, k - a)$ and $(h, k + a)$ then the equation of the ellipse is

$$\frac{(x - h)^2}{b^2} + \frac{(y - k)^2}{a^2} = 1.$$

7.14. THE HYPERBOLA.   When the slicing plane does not pass through V but cuts into both nappes (that is, both "halves") of the cone, the intersection with the cone is called a *hyperbola*. The slicing plane intersects the cone in both of the "halves" of the cone, so that the hyperbola consists of two separate parts. There are two Dandelin spheres. One is in the upper half of the cone with center at S and is tangent to the cone in the circle with center at U and tangent to the slicing plane at F_1. The second Dandelin sphere lies in the lower half of the cone with its center at T and is tangent to the cone in the circle with center at W and to the slicing plane at F_2.

The point P is selected on one of the branches of the hyperbola. The element VP is tangent to Dandelin sphere S at point A and to Dandelin sphere T at B. Since P lies on the slicing plane, PF_1 is tangent to S at F_1, and PF_2 is tangent to T at F_2. It follows that the following line segments are equal:

$$PF_1 = PA,$$
$$PF_2 = PB.$$

The plane of circle U is parallel to the plane of circle W, so

$$PB - PA = AB, \quad \text{and}$$
$$PF_2 - PF_1 = AB = \text{a constant independent of the choice}$$
$$\text{of } P \text{ on a given hyperbola.}$$

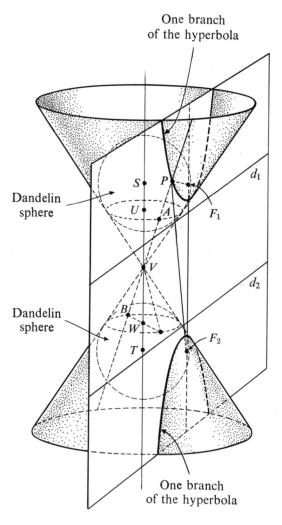

The hyperbola

Thus, we have the fundamental property of the hyperbola:

A hyperbola is the set of all points in a plane that are so located that the difference of the distances from each point to two fixed points is constant.

The two points F_1 and F_2 are called the *foci* of the hyperbola.

Line d_1 in which the slicing plane intersects the plane of circle U and line d_2 in which the slicing plane intersects the plane of circle W are the *directrices* of the hyperbola. The ratio of the distance from a point P to the focus to the distance of P from a directrix is constant for a given hyperbola and is called the *eccentricity* of the hyperbola.

When we replace the slicing plane by a cartesian plane in which the unit distance is chosen and the axes are placed so that the foci are at the points

$$F_1 = (c, 0) \qquad \text{and} \qquad F_2 = (-c, 0), \qquad c > 0,$$

we may derive the equation for the hyperbola. We designate the difference between the distances from a point $P = (x, y)$ to the foci by $2a$, so that we must have, for each point P on the hyperbola,

$$d(P, F_2) - d(P, F_2) = 2a.$$

When this equation is translated into algebraic form, we have

$$\sqrt{(x - c)^2 + y^2} - \sqrt{(x + c)^2 + y^2} = 2a.$$

This reduces by elementary algebraic manipulation to

$$\frac{x^2}{a^2} - \frac{y^2}{(c^2 - a^2)} = 1.$$

If we replace $c^2 - a^2 = b^2$ we obtain the final form of the equation of the hyperbola:

$$\frac{x^2}{a^2} - \frac{y^2}{b^2} = 1.$$

The points $(-a, 0)$ and $(a, 0)$ are the vertices of the hyperbola. The line segment determined by the vertices is called the real axis. The points $(0, -b)$ and $(0, b)$ are the ends of the conjugate axis. The lines $bx - ay = 0$ and $bx + ay = 0$ are called the *asymptotes* of the hyperbola. The hyperbola never touches, or crosses, an asymptote.

The following is the graph of the hyperbola just discussed:

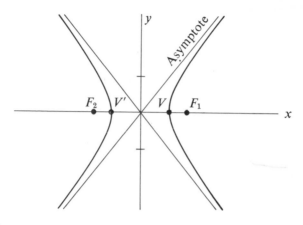

Again, the reader may verify the following facts about hyperbolas:

(a)   If the foci are on the y axis at $(0, -c)$ and $(0, c)$, then the equation of the hyperbola becomes

$$\frac{y^2}{a^2} - \frac{x^2}{b^2} = 1.$$

(b)   If the center is at (h, k) and the foci are at $(h - a, k)$ and $(h + a, k)$, then the equation of the hyperbola becomes

$$\frac{(x - h)^2}{a^2} - \frac{(y - k)^2}{b^2} = 1.$$

(c)   If the center is at (h, k) and the foci are at $(h, k - a)$ and $(h, k + a)$, then the equation of the hyperbola becomes

$$\frac{(y - k)^2}{a^2} - \frac{(x - h)^2}{b^2} = 1.$$

7.15. CARE IN STATING DEFINITIONS.   The geometric definitions of conic sections provide for a good lesson in emphasizing the necessity for being careful in stating a definition. It is very easy to make an incorrect definition by an unintentional omission of a word or phrase or by ignoring possible special cases which should be covered by the definition.

The definition of a *parabola* is frequently given as: "A parabola is

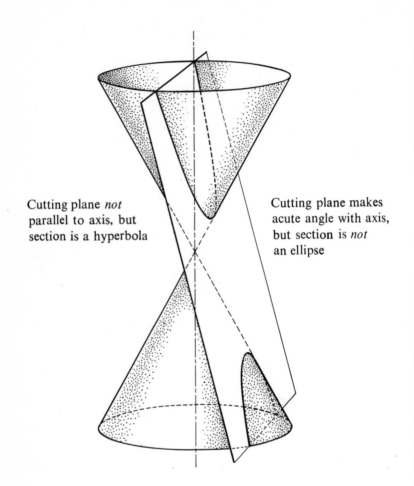

Cutting plane *not* parallel to axis, but section is a hyperbola

Cutting plane makes acute angle with axis, but section is *not* an ellipse

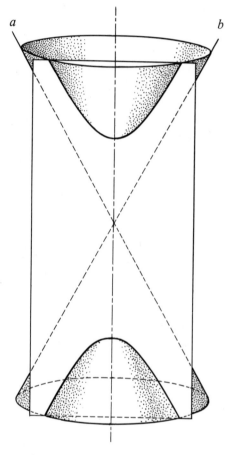

a

b

Section parallel to
elements *a* and *b* of
the cones, but section
is *not* a parabola

the intersection of the cone with a slicing plane parallel to *an* element of the cone and not passing through the vertex of the cone." The necessity for correcting this definition to, "A parabola is the intersection of the cone with a slicing plane parallel to *one and only one element*, but not passing through the vertex of the cone," is made clear in the figure on p. 132.

The definition of the ellipse is usually given as "The ellipse is the intersection of the cone with a slicing plane which intersects the axis of the cone at an acute angle and does not pass through the vertex of the cone". The necessity for stating this definition more precisely as, "The ellipse is the intersection of the cone with a slicing plane which does not pass through the vertex of the cone and which intersects all the elements of one nappe of the cone," is demonstrated in the figure on p. 133.

7.16. A NON-EUCLIDEAN DISTANCE IN THE PLANE.

The distance $d(P_1, P_2)$, which was defined in 7.2.5, is called the *euclidean distance* between points P_1 and P_2. This is the accepted "natural" definition of distance. In this section we shall examine a distance that will satisfy the required properties described in 7.2.1, 7.2.2, 7.2.3, and 7.2.4, but is otherwise very different from the euclidean distance.

Given the points $P_1 = (x_1, y_1)$ and $P_2 = (x_2, y_2)$, we designate a new distance between P_1 and P_2 by $D(P_1, P_2)$, where

$$D(P_1, P_2) = |x_2 - x_1| + |y_2 - y_1|.$$

This distance is the sum of the lengths of two sides of a rectangle for which P_1 and P_2 are opposite vertices, as is shown in the following graph:

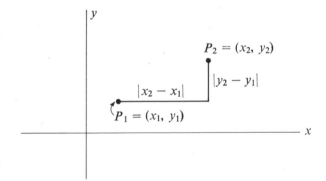

This distance satisfies four conditions for a distance like those listed in 7.2.1 through 7.2.4:

7.16.1. $D(P_1, P_2) \geq 0$;

7.16.2. $D(P_1, P_2) = 0 \leftrightarrow P_2$ coincides with P_1;

7.16.3. $D(P_1, P_2) = D(P_2, P_1)$;

7.16.4. $D(P_1, P_2) + D(P_2, P_3) \geq D(P_1, P_3)$.

Only 7.16.4 requires any proof. If we recall that for every pair of real numbers a, b, we have $|a + b| \leq |a| + |b|$, then

$$\begin{aligned} D(P_1, P_3) &= |x_3 - x_1| + |y_3 - y_1| \\ &= |(x_3 - x_2) + (x_2 - x_1)| + |(y_3 - y_2) + (y_2 - y_1)| \\ &\leq |x_3 - x_2| + |y_3 - y_2| + |x_2 - x_1| + |y_2 - y_1| \\ &= D(P_2, P_3) + D(P_1, P_2). \end{aligned}$$

If we define "conic sections" in the plane with this new non-euclidean distance similar to the way we described them in the preceding sections, we shall obtain some interesting results. In the following we derive the equations and draw the graphs of the non-euclidean conic sections.

7.16.5. *A Non-Euclidean Circle.* A circle is defined as the set of points (x, y) that are at a fixed distance r from a fixed center. If we take the center at $(0, 0)$, then the non-euclidean circle has the equation

$$|x| + |y| = r.$$

In order to see the graph of this non-euclidean circle we must construct it in four pieces:

(a)   When $x \geq 0$ and $y \geq 0$, the equation is $x + y = r$ and the graph is the line segment AB.

(b)   When $x < 0$ and $y \geq 0$, the equation is $-x + y = r$ and graph is the line segment BC.

(c)   When $x < 0$ and $y < 0$, then the equation is $-x - y = r$ and the graph is the line segment CD.

(e)   When $x \geq 0$ and $y < 0$, then the equation is $x - y = r$ and the graph is the line segment DA.

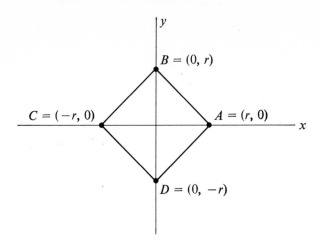

A non-euclidean circle

7.16.6. *A Non-Euclidean Ellipse.* The non-euclidean ellipse is defined as the set of points (x, y) such that the sum of the distances from (x, y) to two fixed foci, $F_1 = (-c, 0)$ and $F_2 = (c, 0)$, is constantly equal to $2a$. The equation of the non-euclidean ellipse is obtained from

$$D(F_1, P) + D(F_2, P) = 2a.$$

This gives us the equation

$$|x - c| + |x + c| + 2|y| = 2a,$$

where $a > c$.

We must graph this non-euclidean ellipse in six segments:

(a) When $x \geq c$, $y \geq 0$, then $x - c \geq 0$ and $x + c \geq 0$, and the equation of the ellipse is

$$(x - c) + (x + c) + 2y = 2a,$$

which reduces easily to

$$x + y = a,$$

for which the graph is the segment AB.

(b) When $-c \leq x \leq c$, $y \geq 0$, then $-c \leq (x - c) \leq 0$ and $0 \leq (x + c) \leq 2c$, and the equation is

$$-(x - c) + (x + c) + 2y = 2a,$$

which reduces to

$$y = a - c$$

for which the graph is the segment BC.

(c) When $x < -c$, $y \geq 0$, then $(x - c) < -2c < 0$ and $(x + c) < 0$, and the equation is

$$-(x - c) - (x + c) + 2y = 2a,$$

which reduces to

$$-x + y = a,$$

for which the graph is the segment CD.

(d) When $x < -c$, $y < 0$, then $(x - c) < -2c < 0$ and $(x + c) < 0$, and the equation is

$$-(x - c) - (x + c) - 2y = 2a,$$

which reduces to

$$-x - y = \mathrm{a},$$

for which the graph is the segment DE.

(e) When $-c \leq x \leq c$, $y < 0$, then $-2c \leq (x - c) \leq 0$ and $0 \leq (x + c) \leq 2c$, and the equation is

$$-(x - c) + (x + c) - 2y = 2a,$$

which reduces to

$$y = -a + c,$$

for which the graph is the line segment EF.

(f) When $x \geq c$, $y < 0$, then $(x - c) \geq 0$ and $(x + c) \geq 2c > 0$ and the equation becomes

$$(x - c) + (x + c) - 2y = 2a,$$

which reduces to

$$x - y = a,$$

for which the graph is the segment FA.

The graph of the non-euclidean ellipse is

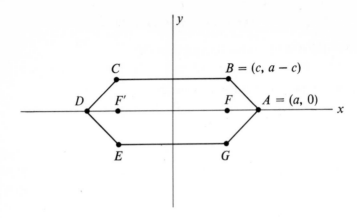

A non-euclidean ellipse

7.16.7. *A Non-Euclidean Parabola.* The non-euclidean parabola will consist of the set of points that are equidistant from the fixed focus, $F = (p, 0)$, and the directrix, $y = -p$, where $p > 0$. The non-euclidean distance, $D(F, P) = |x - p| + |y|$, and the distance from $P = (x, y)$ to the directrix is $|x + p|$. So, the equation of the parabola is

$$|x - p| + |y| = |x + p|, \quad \text{or}$$
$$|y| = |x + p| - |x - p|.$$

We shall construct the graph of this non-euclidean parabola in four segments:

(a) When $x \geq p, y \geq 0$, then $x - p \geq 0$ and $x + p > 0$, and the equation of the non-euclidean parabola is

$$y = (x + p) - (x - p) = 2p,$$

the graph of which is the half line BA.

(b) When $0 \leq x < p, y \geq 0$, then $-p \leq (x - p) < 0$ and $p \leq (x + p) < 2p$, and the equation of the non-euclidean parabola is

$$y = (x + p) + (x - p) = 2x,$$

the graph of which is the line segment BO.

(c) When $0 \leq x < p, y < 0$, then $-p \leq (x - p) < 0$ and $p \leq (x + p) < 2p$, and the equation becomes

$$-y = (x + p) + (x - p) = 2x,$$

the graph of which is the line segment OC.

(d)   When $x > p$, $y < 0$, then $(x - p) > 0$ and $(x + p) > 0$ and the equation becomes

$$-y = (x + p) - (x - p) = 2p,$$

the graph of which is the half line CD.

The other combinations for x and y lead to impossible relations. For one example, if $-p < x < 0$, $y \geq 0$, then $0 < (x + p) < p$ and $-2p < (x - p) < -p$ and the equation is $y = (x + p) + (x - p) = 2x$, for which no points can be found.

The graph of the non-euclidean parabola is:

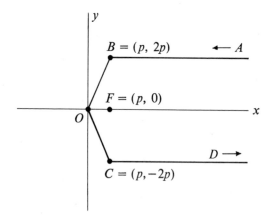

A non-euclidean parabola

PROBLEM SET 7.5

1. Find the equation of each of the following circles:
    a.   with center at $(0, 0)$ and radius of 5;
    b.   with center at $(3, 5)$ and radius of 7;
    c.   with center at $(-3, 6)$ and radius of 9.

2. Find the equation of the parabola with vertex at $(0, 0)$ and focus at $(3, 0)$.

3. Find the equation of the parabola with vertex at $(2, 3)$ and directrix in the line $x = -4$.

4. Find the equation of the ellipse with center at $(0, 0)$ and with major axis 12 units long and minor axis 8 units long.

5. Find the equation of the ellipse with foci at $(6, 0)$ and $(-6, 0)$ and minor axis 16 units long.

6. Find the equation of the hyperbola with foci at $(10, 0)$ and $(-10, 0)$ and vertices at $(6, 0)$ and $(-6, 0)$.

7. Find the equation of the hyperbola with center at $(0, 0)$, one focus at $(6, 0)$, and one asymptote in the line $2y = x\sqrt{5}$.

8. Find the non-euclidean distance $D(A, B)$ if
    a. $A = (5, -7)$ and $B = (-1, -11)$;
    b. $A = (6, 2)$ and $B = (15, -9)$.

9. Find the equation of the non-euclidean circle with center at $(0, 0)$ and radius of 3. Sketch the graph.

10. Find the equation of the non-euclidean circle with center at $(2, 3)$ and radius of 8. Sketch the graph.

11. Find the equation of the non-euclidean ellipse with foci at $(4, 0)$ and $(-4, 0)$ and with $a = 5$. Sketch the graph.

12. Find the equation of the non-euclidean ellipse with foci at $(10, 0)$ and $(-10, 0)$ and with the ends of the major axis at $(15, 0)$ and $(-15, 0)$. Sketch the graph.

13. Find the equation of the non-euclidean parabola with focus at $(5, 0)$ and with vertex at the origin. Sketch the graph.

14. Investigate the possibility of having a non-euclidean hyperbola.

15. Experiment with the invention of other non-euclidean distances and, if you find some, investigate equations of conic sections, and so on.

16. In the drawing of the ellipse in Section 7.13 develop the ratio defined as the eccentricity. Show that in the ellipse the eccentricity must be less than 1. Find the equations of the directrices in the cartesian plane of the ellipse.

17. In the drawing of the hyperbola in Section 7.14 develop the ratio defined as the eccentricity. Show that in the hyperbola the eccentricity is greater than 1. Find the equations of the directrices, and the eccentricity in the cartesian plane for the hyperbola.

18. Prove that the eccentricity for the parabola is equal to 1.

CHAPTER EIGHT

Analytic Geometry in 3-Space

8.1. CARTESIAN 3-SPACE. We have three cartesian lines, x, y, z, with the same unit distance. We may establish three cartesian planes: the x-y plane, the x-z plane, and the y-z plane. If these three planes are placed in 3-space so that they intersect on the cartesian lines x, y, z, and the real number 0 is at the intersection of all three planes, so that each plane is perpendicular to the other two planes, then we have established cartesian 3-space. The three cartesian lines and the three cartesian planes form a frame of reference in 3-space by means of which each point may correspond to an ordered triple of real numbers, and, conversely, for each ordered triple of real numbers there is a point in cartesian 3-space. The set

$$K_3 = K \times K \times K = \{(x, y, z) \; ; \; x, y, z \in K\}$$

of all ordered triples of real numbers will also be called the set of all points in cartesian 3-space.

8.2. DISTANCE BETWEEN TWO POINTS IN 3-SPACE. For each pair of points $P_1 = (x_1, y_1, z_1)$ and $P_2 = (x_2, y_2, z_2)$ there is a nonnegative real number, designated by the symbol $d(P_1, P_2)$ and called the distance between P_1 and P_2, having the following properties:

8.2.1. $d(P_1, P_2) \geq 0$;

8.2.2. $d(P_1, P_2) = 0 \leftrightarrow P_2$ coincides with P_1;

8.2.3. $d(P_1, P_2) = d(P_2, P_1)$;

8.2.4. $d(P_1, P_2) + d(P_2, P_3) \geq d(P_1, P_3)$;

8.2.5. $d(P_1, P_2) = \sqrt{(x_2 - x_1)^2 + (y_2 - y_1)^2 + (z_2 - z_1)^2}.$

The distance $d(P_1, P_2)$ is also called the *length* of the line segment P_1P_2.

8.3. VECTOR.   A line segment, AB, connecting points A and B and having the length of $d(A, B)$, has the same *direction* as the line AB which contains the line segment AB. We shall say a little more later about the description of the direction of a line in 3-space. If it should make a difference whether one goes from A to B, or from B to A on the line segment AB, then the line segment is said to have *sense*. Sense is designated as either positive or negative. If going from A to B is the *positive sense*, then A is called the *initial point* and B is called the *terminal point*. The sense of a line segment is designated by an arrowhead at its terminal point. If the sense is positive from A to B, then the sense is *negative* from B to A. Thus, a sketch of the line segment AB with its initial point at A and terminal point at B and with positive sense from A to B is as follows:

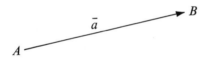

The set of all line segments with the same length, direction, and sense as a given line segment, AB, is called the *vector AB*. The vector AB as just described will be designated by the symbol $\overrightarrow{AB}$, or by single letter symbols, like $\bar{a}, \bar{b}, \bar{c}, \ldots$. We may use the single letter symbol only when the three essential properties of length, direction, and sense are clearly defined.

The length of a vector is designated by using the absolute value symbol on the symbol for the vector. Thus, vector lengths are designated by the notation $|\overrightarrow{AB}|, |\bar{a}|, |\bar{b}|, |\bar{c}|$, and so on. Then,

8.3.1. $|\overrightarrow{AB}| = |\overleftarrow{AB}| = d(A, B).$

If a vector is multiplied by a scalar, K, the result is a vector with the same direction, the same sense if the scalar is positive but opposite sense if the scalar is negative, and with its length multiplied by $|k|$. This result is summarized in the following:

8.3.2.   If $k \in K$ is a scalar and $k \neq 0$, then the vector $k \overrightarrow{AB}$, or $k\bar{a}$, is a vector such that

a. $|k \overrightarrow{AB}| = |k| \, |\overrightarrow{AB}| = |k| \, d(A, B)$,   or
$|k\bar{a}| = |k| \, |\bar{a}|$.

b.   The direction of $k \overrightarrow{AB}$, or of $k\bar{a}$, is the same direction as that of $\overrightarrow{AB}$, or of $\bar{a}$.

c.   If $k > 0$, then the sense of $k \overrightarrow{AB}$, or of $k\bar{a}$, is the same as the sense of $\overrightarrow{AB}$, or of $\bar{a}$;

if $k < 0$, then the sense of $k \overrightarrow{AB}$, or of $k\bar{a}$, is opposite to the sense of $\overrightarrow{AB}$, or of $\bar{a}$.

As a consequence of 8.3.2.c we have

$$\overleftarrow{AB} = -1 \cdot \overrightarrow{AB} = -\overrightarrow{AB}.$$

8.4. ADDITION OF TWO VECTORS.   Since a vector is defined to be a set of line segments in which each element has a certain specified length, direction, and sense, it follows that each vector $\bar{a}$ is a representative notation for a set of line segments. The actual position of vector $\bar{a}$ in 3-space is meaningless. In fact, vector $\bar{a}$ is actually everywhere in 3-space.

The *vector sum* of two vectors $\bar{a}$ and $\bar{b}$ is obtained by finding the vector $\bar{b}$ which has its initial point at the terminal point of $\bar{a}$. Then the vector sum, $\bar{a} + \bar{b}$, is the vector from the initial point of $\bar{a}$ to the terminal point of $\bar{b}$. The vector sum is sketched in the following:

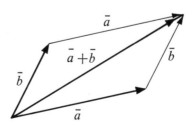

Since the two vectors, $\bar{a}$ and $\bar{b}$, determine a unique parallelogram with the two given vectors as adjacent sides, the vector $\bar{a} + \bar{b}$ is a diagonal of this parallelogram. It follows at once from the definition and from the sketch that vector addition is commutative, that is,

$$\bar{b} + \bar{a} = \bar{a} + \bar{b}.$$

From the definition of $-\bar{a}$ in Section 8.3 it follows that

$$\bar{a} + (-\bar{b}) = \bar{a} - \bar{b}$$

is the diagonal of the parallelogram other than the one which is the vector sum. This is shown in the following sketch:

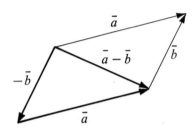

Since vectors $\bar{a}$, $\bar{b}$, and $\bar{a} + \bar{b}$ form the sides of a triangle, and, likewise, $\bar{a}$, $(-\bar{b})$, and $\bar{a} - \bar{b}$ form the sides of another triangle, we have that

8.4.1. $|\bar{a} + \bar{b}| \leq |\bar{a}| + |\bar{b}|$;

8.4.2. $|\bar{a} - \bar{b}| \leq |\bar{a}| + |\bar{b}|$;

8.4.3. $|\bar{a} - \bar{b}| \geq ||\bar{a}| - |\bar{b}||$.

The vector sum, $\bar{a} + \bar{b}$, is frequently called the *resultant* of $\bar{a}$ and $\bar{b}$, particularly in the applications to problems in physics.

From the use of theorems on similar triangles the reader can easily verify each of the following:

8.4.4. $k\bar{a} + k\bar{b} = k(\bar{a} + \bar{b})$, for $k \in K$, $k \neq 0$.

8.4.5. $k\bar{a} - k\bar{b} = k(\bar{a} - \bar{b})$, for $k \in K$, $k \neq 0$.

For three vectors, $\bar{a}$, $\bar{b}$, $\bar{c}$, the following associative properties are valid:

8.4.6. $\bar{a} + (\bar{b} + \bar{c}) = (\bar{a} + \bar{b}) + \bar{c};$

8.4.7. $\bar{a} + (\bar{b} - \bar{c}) = (\bar{a} + \bar{b}) - \bar{c}.$

A vector whose length is zero is called a *zero vector*. A zero vector is denoted by $\bar{0}$, where $|\bar{0}| = 0$. A zero vector is an additive identity, since, for every $\bar{a}$,

$$\bar{a} + \bar{0} = \bar{a}.$$

The inverse for $\bar{a}$ with respect to vector addition is the vector $(-\bar{a})$, since

$$\bar{a} + (-\bar{a}) = \bar{0}.$$

A vector whose length is 1 is called a *unit vector*. If $\bar{u}$ is a unit vector, then $|\bar{u}| = 1$. For every nonzero vector, $\bar{a}$, there is a unit vector, $\bar{u}$, such that $\bar{a} = |\bar{a}|\bar{u}$. The vector $\bar{u}$ is called the *normal vector* for $\bar{a}$, and

$$\bar{u} = \frac{\bar{a}}{|\bar{a}|}.$$

8.5. COORDINATES OF VECTORS.   When vector $\bar{a}$ is placed in cartesian 3-space we choose the member of the set that has its initial point at the origin, $0 = (0, 0, 0)$, and its terminal point at the point $A = (a_1, a_2, a_3)$. Then, the length of $\bar{a}$ is

$$|\bar{a}| = \sqrt{a_1^2 + a_2^2 + a_3^2}.$$

The direction of $\bar{a}$ is that of the line determined by the points 0 and A, and the sense of $\bar{a}$ is from 0 to A. The coordinates of the terminal point, A, are also the coordinates of the vector $\bar{a}$, so that we write

$$\bar{a} = (a_1, a_2, a_3).$$

If $\bar{b} = (b_1, b_2, b_3)$ then the reader may prove by simple geometric developments that

$$\bar{a} + \bar{b} = (a_1 + b_1, a_2 + b_2, a_3 + b_3), \quad \text{and}$$
$$\bar{a} - \bar{b} = (a_1 - b_1, a_2 - b_2, a_3 - b_3).$$

If $A = (a_1, a_2, a_3)$ and $B = (b_1, b_2, b_3)$ are two distinct points, then the vector $\overrightarrow{AB}$ is the vector $(\bar{b} - \bar{a})$, and we have

$$\overrightarrow{AB} = (\bar{b} - \bar{a}) = (b_1 - a_1, b_2 - a_2, b_3 - a_3), \quad \text{and}$$

$$\overleftarrow{AB} = (\bar{a} - \bar{b}) = (a_1 - b_1, a_2 - b_2, a_3 - b_3).$$

We are in the process of constructing an algebra of vectors. Up to this point either we have established or the reader can easily establish each of the following:

8.5.1. *Equality.* $\bar{a} = \bar{b} \leftrightarrow a_1 = b_1$, $a_2 = b_2$, and $a_3 = b_3$; that is, equality means identity in the sense that two vectors are equal if and only if they are actually the same vector.

8.5.2. If k is a scalar, $k \in K$, then

$$k\bar{a} = k(a_1, a_2, a_3) = (ka_1, ka_2, ka_3).$$

8.5.3. *Addition.* $\bar{a} + \bar{b} = (a_1 + b_1, a_2 + b_2, a_3 + b_3)$.

8.5.4. *Subtraction.* $\bar{a} - \bar{b} = (a_1 - b_1, a_2 - b_2, a_3 - b_3)$.

8.5.5. *Commutative Law for Addition.* $\bar{a} + \bar{b} = \bar{b} + \bar{a}$; since the coordinates are real numbers it follows that

$$(a_1 + b_1, a_2 + b_2, a_3 + b_3) = (b_1 + a_1, b_2 + a_2, b_3 + a_3).$$

8.5.6. *Associative Law for Addition.* $\bar{a} + (\bar{b} + \bar{c}) = (\bar{a} + \bar{b}) + \bar{c}$, since

$$(a_1 + (b_1 + c_1), a_2 + (b_2 + c_2), a_3 + (b_3 + c_3))$$
$$= ((a_1 + b_1) + c_1, (a_2 + b_2) + c_2, (a_3 + b_3) + c_3).$$

8.5.7. *The Zero Vector is the Additive Identity.* $\bar{0} = (0, 0, 0)$, and for every vector $\bar{a} = (a_1, a_2, a_3)$, $\bar{a} + \bar{0} = \bar{a}$.

8.5.8. *Additive Inverse Vectors.* The additive inverse of $\bar{a} = (a_1, a_2, a_3)$ is the vector $(-\bar{a}) = (-a_1, -a_2, -a_3)$ since $\bar{a} + (-\bar{a}) = \bar{0}$.

8.5.9. *Unit Vectors.* If $\bar{u}$ is a unit vector, then $\bar{u} = \sqrt{u_1^2 + u_2^2 + u_3^2} = 1$. The *normal vector* for $\bar{a}$ is the vector $\bar{u}$, where

$$\bar{u} = \frac{1}{|\bar{a}|} (a_1, a_2, a_3) = \left(\frac{a_1}{|\bar{a}|}, \frac{a_2}{|\bar{a}|}, \frac{a_3}{|\bar{a}|} \right).$$

8.5.10. *The Fundamental Unit Vectors.* The three vectors $\bar{i}, \bar{j}, \bar{k}$, are called the fundamental unit vectors, and

$$\bar{i} = (1, 0, 0);$$
$$\bar{j} = (0, 1, 0);$$
$$\bar{k} = (0, 0, 1).$$

8.5.11. *Components of Vectors.* For every vector $\bar{a}$ it is possible to write $\bar{a}$ as a combination of the fundamental unit vectors, $\bar{i}, \bar{j}, \bar{k}$, in the following way: $\bar{a} = (a_1, a_2, a_3) = a_1\bar{i} + a_2\bar{j} + a_3\bar{k}$, where the vectors $a_1\bar{i}, a_2\bar{j}, a_3\bar{k}$ are the *component vectors* for $\bar{a}$.

It should be noted that the fundamental unit vectors $\bar{i}, \bar{j}, \bar{k}$ lie on the x, y, z axes, respectively. The components of $\bar{a}$ are vectors which lie on, or are parallel to, the coordinate axes.

Since the coordinates of a vector determine each of the three properties of length, direction, and sense, we may now define the direction of a line to be the direction of a vector that lies on it. Thus, the line determined by the two points $A = (a_1, a_2, a_3)$ and $B = (b_1, b_2, b_3)$ contains the vector $(b_1 - a_1, b_2 - a_2, b_3 - a_3)$. The direction of the line is the direction of this vector. The coordinates of the vector lying on the line are called the *direction numbers* of the line. The normal vector for a vector that lies on a line is an ordered triple of numbers which are called the *direction cosines* of the line.

For an example, the line determined by the points $A = (2, -1, 5)$ and $B = (5, 3, -7)$ contains the vector $\bar{b} - \bar{a} = (3, 4, -12)$. The direction numbers of line AB are $(3, 4, -12)$. The direction cosines of line AB are $(3/13, 4/13, -12/13)$.

If a line a contains the vector $\bar{a}$, then any other line which contains vector $\bar{a}$, or a vector $k\bar{a}$, is parallel to line a.

8.6. THE DOT, OR SCALAR, PRODUCT OF TWO VECTORS.

Given two vectors $\bar{a}$ and $\bar{b}$ with their initial points coincident. If we construct a perpendicular from the terminal point of $\bar{b}$ to $\bar{a}$, or to the line which contains $\bar{a}$, then the vector from the initial point of $\bar{a}$ to the foot of this perpendicular is called the projection of $\bar{b}$ upon $\bar{a}$, which is

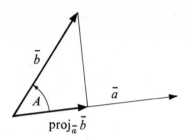

designated by $\text{proj}_{\bar{a}}\bar{b}$. We designate the angle measure from $\bar{a}$ to $\bar{b}$ by A. Then the vector, $\text{proj}_{\bar{a}}\bar{b}$, has the same direction and sense as $\bar{a}$, but its length is

$$|\text{proj}_{\bar{a}}\bar{b}| = |\bar{b}| \cos A, \qquad -\pi/2 \leq A \leq \pi/2.$$

If $\pi/2 \leq A \leq 3\pi/2$, then the sense of $\text{proj}_{\bar{a}}\bar{b}$ is opposite to the sense of $\bar{a}$.

The *dot*, or *scalar*, *product* $\bar{a} \cdot \bar{b}$ is defined to be the product of the length of $\bar{a}$ by the length of $\text{proj}_{\bar{a}}\bar{b}$, that is:

8.6.1. *Dot, or Scalar, Product.*

$$\bar{a} \cdot \bar{b} = |\bar{a}| \, |\text{proj}_{\bar{a}}\bar{b}| = |a| \, |b| \cos A,$$

with the added provision that the dot product is positive if $\cos A$ is positive, and it is negative when $\cos A$ is negative. Thus,

$$\begin{array}{lll}
\bar{a} \cdot \bar{b} > 0, & \text{if} & -\pi/2 < A < \pi/2; \\
\bar{a} \cdot \bar{b} < 0, & \text{if} & \pi/2 < A < 3\pi/2; \\
\bar{a} \cdot \bar{b} = 0, & \text{if} & A = \pi/2.
\end{array}$$

According to the last statement, two nonzero vectors, $\bar{a}$ and $\bar{b}$, are perpendicular if and only if the dot product $\bar{a} \cdot \bar{b} = 0$.

The following properties of the dot product follow almost immediately from the definition:

8.6.2. *Commutative Law.* $\bar{b} \cdot \bar{a} = \bar{a} \cdot \bar{b}$.

8.6.3. The dot product of a vector with itself is the square of its length: $\bar{a} \cdot \bar{a} = |\bar{a}|^2$.

8.6.4. For a scalar, $k \in K$, $(k\bar{a}) \cdot \bar{b} = \bar{a} \cdot (k\bar{b}) = k(\bar{a} \cdot \bar{b})$.

8.6.5. The associative law has no meaning in connection with the dot product.

8.6.6. The dot product is distributive over addition:

$$\bar{a} \cdot (\bar{b} + \bar{c}) = (\bar{a} \cdot \bar{b}) + (\bar{a} \cdot \bar{c}),$$

and

$$\bar{a} \cdot (p\bar{b} + q\bar{c}) = p(\bar{a} \cdot \bar{b}) + q(\bar{a} \cdot \bar{c}), \text{ where } p, q \in K$$

are scalars.

8.6.7. Since the fundamental unit vectors, $\bar{i}, \bar{j}, \bar{k}$, are mutually perpendicular, it follows that

$$\bar{i} \cdot \bar{j} = \bar{j} \cdot \bar{k} = \bar{k} \cdot \bar{i} = 0.$$

When we express each of $\bar{a}$ and $\bar{b}$ in terms of its components, we have

8.6.8. $$\bar{a} \cdot \bar{b} = (a_1\bar{i} + a_2\bar{j} + a_3\bar{k}) \cdot (b_1\bar{i} + b_2\bar{j} + b_3\bar{k})$$
$$= a_1b_1 + a_2b_2 + a_3b_3.$$

If $\bar{a}$ is perpendicular to $\bar{b}$, then $(\bar{a} + \bar{b})$ is the hypotenuse of a right triangle for which $\bar{a}$ and $\bar{b}$ are the sides, and $\bar{a} \cdot \bar{b} = 0$. In this right triangle we have

$$|\bar{a} + \bar{b}|^2 = (\bar{a} + \bar{b}) \cdot (\bar{a} + \bar{b})$$
$$= (\bar{a} \cdot \bar{a}) + (\bar{a} \cdot \bar{b}) + (\bar{b} \cdot \bar{a}) + (\bar{b} \cdot \bar{b})$$
$$= |\bar{a}|^2 + |\bar{b}|^2,$$

which is the well-known pythagorean theorem for right triangles.

In a parallelogram with adjacent sides $\bar{a}$ and $\bar{b}$ the diagonals are $(\bar{a} + \bar{b})$ and $(\bar{a} - \bar{b})$. The dot product of the diagonals is

$$(\bar{a} - \bar{b}) \cdot (\bar{a} + \bar{b}) = (\bar{a} \cdot \bar{a}) - (\bar{b} \cdot \bar{b}) = |\bar{a}|^2 - |\bar{b}|^2$$
$$= (|\bar{a}| - |\bar{b}|)(|\bar{a}| + |\bar{b}|).$$

Moreover, if the diagonals are perpendicular, then we have

$$(\bar{a} - \bar{b}) \cdot (\bar{a} + \bar{b}) = |\bar{a}|^2 - |\bar{b}|^2 = 0,$$

so that $|\bar{a}| = |\bar{b}|$. In other words, if the diagonals of a parallelogram are perpendicular, then the sides are equal, so that the parallelogram is a rhombus, or a square.

Again, in the same parallelogram,

$$\begin{aligned} |\bar{a} - \bar{b}|^2 + |\bar{a} + \bar{b}|^2 &= (\bar{a} - \bar{b}) \cdot (\bar{a} - \bar{b}) + (\bar{a} + \bar{b}) \cdot (\bar{a} + \bar{b}) \\ &= 2(\bar{a} \cdot \bar{a}) + 2(\bar{b} \cdot \bar{b}) \\ &= 2|\bar{a}|^2 + 2|\bar{b}|^2, \end{aligned}$$

which says "the sum of the squares of the lengths of the diagonals of a parallelogram is equal to the sum of the squares of the lengths of the sides."

Since the vectors $\bar{a}$, $\bar{b}$, $(\bar{b} - \bar{a})$ form the sides of a triangle, we have

$$\begin{aligned} |\bar{b} - \bar{a}|^2 &= (\bar{b} - \bar{a}) \cdot (\bar{b} - \bar{a}) \\ &= (\bar{b} \cdot \bar{b}) + (\bar{a} \cdot \bar{a}) - 2(\bar{a} \cdot \bar{b}) \\ &= |b|^2 + |a|^2 - 2|b|\,|a| \cos A, \end{aligned}$$

which we recognize as the well-known law of cosines in trigonometry.

From the definition of the dot product we obtain

$$\cos A = \frac{\bar{a} \cdot \bar{b}}{|\bar{a}|\,|\bar{b}|} \qquad \text{and} \qquad \cos^2 A = \frac{(\bar{a} \cdot \bar{b})^2}{|\bar{a}|^2|\bar{b}|^2},$$

and from this we further obtain

$$\sin^2 A = 1 - \cos^2 A = \frac{|\bar{a}|^2|\bar{b}|^2 - (\bar{a} \cdot \bar{b})^2}{|\bar{a}|^2|\bar{b}|^2}.$$

8.7. A DETERMINANT OF ORDER 3.

We defined a determinant of order 2 in Section 5.6. Any square array of numbers with a rule for determining its numerical value is called a determinant. The number of rows and columns is the *order* of the determinant.

We are interested in particular determinants of order 3. The first row will be the coordinates of the vector $\bar{a}$, the second row will be the coordinates of the vector $\bar{b}$, and the third row will be the coordinates of the vector $\bar{c}$. This determinant will be designated by the symbol $[\bar{a}, \bar{b}, \bar{c}]$, where

$$[\bar{a}, \bar{b}, \bar{c}] = \begin{vmatrix} a_1 & a_2 & a_3 \\ b_1 & b_2 & b_3 \\ c_1 & c_2 & c_3 \end{vmatrix}.$$

The numerical value of this determinant of order 3 is defined to be

$$[\bar{a}, \bar{b}, \bar{c}] = a_1(b_2c_3 - b_3c_2) - a_2(b_1c_3 - b_3c_1) + a_3(b_1c_2 - b_2c_1)$$
$$= c_1(a_2b_3 - a_3b_2) - c_2(a_1b_3 - a_3b_1) + c_3(a_1b_2 - a_2b_1).$$

8.8. THE CROSS, OR VECTOR, PRODUCT. The *cross product vector*

$$\bar{a} \times \bar{b} = \bar{c}$$

is defined to be a vector $\bar{c}$, such that:

(a) $|\bar{c}| = |\bar{a}| \, |\bar{b}| \, |\sin A|$, where A is the angle measured from $\bar{a}$ to $\bar{b}$;

(b) The direction of $\bar{c}$ is defined by

$$\bar{a} \cdot \bar{c} = \bar{b} \cdot \bar{c} = 0,$$

so that $\bar{c}$ is perpendicular to both $\bar{a}$ and $\bar{b}$ and, consequently, $\bar{c}$ is perpendicular to the plane containing $\bar{a}$ and $\bar{b}$.

(c) The sense is determined by the requirement that $[\bar{a}, \bar{b}, \bar{c}] > 0$.
For an example, we determine the vector (u_1, u_2, u_3) such that

$$\bar{i} \times \bar{j} = (u_1, u_2, u_3).$$

Since $\bar{i}$ is perpendicular to $\bar{j}$ the angle from i to j is $\pi/2$, so the length of the vector

$$\bar{i} \times \bar{j} = 1 \cdot 1 \cdot \sin \pi/2 = 1 = \sqrt{u_1^2 + u_2^2 + u_3^2}.$$

From property (b) we have that

$$(\bar{i} \times \bar{j}) \cdot \bar{i} = (u_1, u_2, u_3) \cdot (1, 0, 0) = u_1 = 0, \quad \text{and}$$
$$(\bar{i} \times \bar{j}) \cdot \bar{j} = (u_1, u_2, u_3) \cdot (0, 1, 0) = u_2 = 0.$$

We now have that $(\bar{i} \times \bar{j}) = (0, 0, u_3)$, so $u_3^2 = 1$, and $u_3 = +1$, or $u_3 = -1$. But property (c) requires that the determinant $[\bar{i}, \bar{j}, (\bar{i} \times \bar{j})]$ be positive. That is,

$$[\bar{i}, \bar{j}, (\bar{i} \times \bar{j})] = \begin{vmatrix} 1 & 0 & 0 \\ 0 & 1 & 0 \\ 0 & 0 & u_3 \end{vmatrix} = u_3 > 0,$$

and since u_3 is either $+1$ or -1, we now see that $u_3 = +1$, and

$$\bar{i} \times \bar{j} = (0, 0, 1) = k.$$

We may obtain in a similar manner that

$$\bar{j} \times \bar{k} = \bar{i},$$
$$\bar{k} \times \bar{i} = \bar{j},$$

$$\bar{j} \times \bar{i} = -\bar{k},$$
$$\bar{k} \times \bar{j} = -\bar{i}, \quad \text{and}$$
$$\bar{i} \times \bar{k} = -\bar{j}.$$

The following properties of the cross product follow almost immediately from the definition:

8.8.1.   For every vector $\bar{a}$, $\bar{a} \times \bar{a} = \bar{0}$.

8.8.2.   For a scalar, $k \in K$, $(k\bar{a}) \times \bar{b} = \bar{a} \times (k\bar{b}) = k(\bar{a} \times \bar{b})$.

8.8.3.   The cross product is not commutative:

$$\bar{a} \times \bar{b} = -(\bar{b} \times \bar{a}).$$

8.8.4.   The cross product is distributive over vector addition:

$$\bar{a} \times (\bar{b} + \bar{c}) = (\bar{a} \times \bar{b}) + (\bar{a} + \bar{c}).$$

If we express each of the factors in component form we can discover a workable procedure for calculating the cross product vector.

$$\bar{a} \times \bar{b} = (a_1\bar{i} + a_2\bar{j} + a_3\bar{k}) \times (b_1\bar{i} + b_2\bar{j} + b_3\bar{k})$$
$$= (a_2b_3 - a_3b_2)\bar{i} - (a_1b_3 - a_3b_1)\bar{j} + (a_1b_2 - a_2b_1)\bar{k}$$
$$= \left(\begin{vmatrix} a_2 & a_3 \\ b_2 & b_3 \end{vmatrix}, \quad -\begin{vmatrix} a_1 & a_3 \\ b_1 & b_3 \end{vmatrix}, \quad \begin{vmatrix} a_1 & a_2 \\ b_1 & b_2 \end{vmatrix} \right).$$

The coordinates of the vector in the last member are expressed as determinants of order 2, which were defined in Section 5.6. These coordinates in determinant form may be easily computed by writing the 2×3 matrix which is formed by writing the coordinates of the first factor in a cross product in the first row and the coordinates of the second factor in the second row. Thus, for the cross product $\bar{a} \times \bar{b}$ we write the matrix

$$\begin{pmatrix} a_1 & a_2 & a_3 \\ b_1 & b_2 & b_3 \end{pmatrix}.$$

Then, the first coordinate in $\bar{a} \times \bar{b}$ is the determinant of order 2 obtained from this matrix by omitting the first column, the second coordinate is the negative of the determinant of order 2 obtained by omitting the second column in the matrix, and the third coordinate is the determinant of order 2 obtained by omitting the third column from the matrix.

We see, by comparing the definition of the value of the determinant of order 3 and the definition of the cross product vector that

$$[\bar{a}, \bar{b}, \bar{c}] = \bar{a} \cdot (\bar{b} \times \bar{c}).$$

8.9. AREAS AND VOLUMES IN 3-SPACE. According to the definition of $\bar{a} \times \bar{b}$, the length of the cross product vector is

$$|\bar{a} \times \bar{b}| = |\bar{a}| \, |\bar{b}| \, |\sin A| = |\bar{a}| \, |\bar{h}|,$$

where $\bar{h}$ is the vector from the terminal point of $\bar{b}$ to $\bar{a}$ and perpendicular to $\bar{a}$, so that $|\bar{h}| = |\bar{b}| \, |\sin A|$. Since the two vectors $\bar{a}$ and $\bar{b}$ determine a

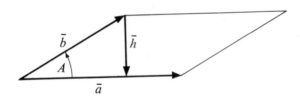

unique parallelogram with them as adjacent sides, the length of the cross product vector, $|\bar{a} \times \bar{b}|$, is the area of this parallelogram.

Since each of the triangles with sides (1) $\bar{a}, \bar{b}, (\bar{a} + b)$ or (2) $\bar{a}, \bar{b},$ $(\bar{b} - \bar{a})$ is one half of the parallelogram, the area of either triangle is $\frac{1}{2}|\bar{a} \times \bar{b}|$.

For an example, we have the three points

$$A = (3, -1, 5), \quad B = (7, 2, -3), \quad C = (-4, 3, 2).$$

The three points, A, B, C are the vertices of a triangle, or three of the vertices of a parallelogram, the fourth vertex of which can be calculated.

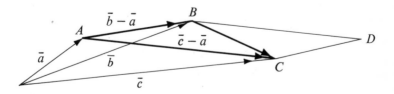

We must first obtain the vectors on the sides of *ABC*. The vectors to the vertices are, respectively,

$$\bar{a} = (3, -1, 5), \quad \bar{b} = (7, 2, -3), \quad \text{and} \quad \bar{c} = (-4, 3, 2).$$

The vectors on the sides *AB* and *AC*, each with its initial point at *A*, are

$$(\bar{b} - \bar{a}) = (4, 3, -8) \quad \text{and}$$
$$(\bar{c} - \bar{a}) = (-7, 4, -3), \quad \text{respectively.}$$

The fourth vertex, *D*, has the coordinates

$$D = (\bar{b} - \bar{a}) + (\bar{c} - \bar{a}) = (-3, 7, -11).$$

The area of the parallelogram *ABCD* is

$$|(\bar{b} - \bar{a}) \times (\bar{c} - \bar{a})| = |(23, 68, 37)| = \sqrt{6522} = 80.759.$$

The area of triangle *ABC* is one half of the area of the parallelogram, 40.379.

Given three noncoplanar vectors $\bar{a}, \bar{b}, \bar{c}$ with their initial points coincident. They determine a unique parallelopiped for which the three given vectors are adjacent sides.

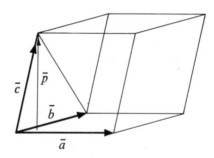

From the definition 8.6.1 of the dot product, we have

$$|(\bar{a} \times \bar{b}) \cdot \bar{c}| = |(\bar{a} \times \bar{b})| \, |\bar{p}|,$$

where $\bar{p}$ is the projection of $\bar{c}$ upon the vector $(\bar{a} \times \bar{b})$. Since the cross product vector $(\bar{a} \times \bar{b})$ is perpendicular to the plane determined by the vectors $\bar{a}$ and $\bar{b}$, then $\bar{p}$ is a vector from the terminal point of $\bar{c}$ to the plane determined by $\bar{a}$ and $\bar{b}$ and $\bar{p}$ is perpendicular to this plane. Consequently, $|\bar{p}|$ is the altitude of the parallelopiped determined by $\bar{a}, \bar{b}, \bar{c}$

from the terminal point of $\bar{c}$ to the base. Since $|\bar{a} \times \bar{b}|$ is the area of the base of the parallelopiped, it follows that 8.9.1 is the volume of the parallelopiped.

It follows that

$$|\bar{a}\cdot(\bar{b} \times \bar{c})| = |\bar{b}\cdot(\bar{c} \times \bar{a})| = |\bar{c}\cdot(\bar{a} \times \bar{b})|,$$

since each expression is the volume of the same parallelopiped.

The three adjacent edges $\bar{a}$, $\bar{b}$, $\bar{c}$ also determine a tetrahedron. The volume of this tetrahedron is one sixth of the volume of the parallelopiped, that is,

$$\tfrac{1}{6}|\bar{a}\cdot(\bar{b} \times \bar{c})| = \text{volume of the tetrahedron determined}$$
$$\text{by } \bar{a}, \bar{b}, \bar{c} \text{ as adjacent edges.}$$

From the definition of the value of the determinant of order 3 in Section 8.8, we have that

$$|[\bar{a}\cdot(\bar{b} \times \bar{c})]| = \text{volume of the parallelopiped determined}$$
$$\text{by } \bar{a}, \bar{b}, \bar{c} \text{ as adjacent edges.}$$

For an example, we study the parallelopiped determined by the four points, as vertices:

$$A = (16, 6, 4), \qquad B = (-1, 2, -3),$$
$$C = (1, -1, 3), \qquad D = (4, 9, 7).$$

The following vectors have initial points at the origin and terminal points at the vertices, respectively:

$$\bar{a} = (16, 6, 4), \qquad \bar{b} = (-1, 2, -3),$$
$$\bar{c} = (1, -1, 3), \qquad \bar{d} = (4, 9, 7).$$

The following vectors lie on the edges of the parallelopiped and each has its initial point at A:

$$(\bar{b} - \bar{a}) = (-17, -4, -7),$$
$$(\bar{c} - \bar{a}) = (-15, -7, -1),$$
$$(\bar{d} - \bar{a}) = (-12, 3, 3).$$

The volume of the parallelopiped determined by the three concurrent edges $(\bar{b} - \bar{a})$, $(\bar{c} - \bar{a})$, and $(\bar{d} - \bar{a})$ is

$$V = |(\bar{b} - \bar{a}) \cdot [(\bar{c} - \bar{a}) \times (\bar{d} - \bar{a})]|$$
$$= |(-17, -4, -7) \cdot (-18, 57, -129)|$$
$$= |306 - 228 + 903| = |981| = 981.$$

The remaining four vertices of the parallelopiped are

$$E = \bar{e} = (\bar{c} - \bar{a}) + (\bar{b} - \bar{a}) = (-32, -11, -8),$$
$$F = \bar{f} = (\bar{c} - \bar{a}) + (\bar{d} - \bar{a}) = (-27, -4, 2),$$
$$G = \bar{g} = (\bar{d} - \bar{a}) + (\bar{b} - \bar{a}) = (-29, -1, -4),$$
$$H = \bar{h} = (\bar{f} - \bar{d}) + (\bar{g} - \bar{d}) = (-31, -13, -5) + (-33, -10, -11)$$
$$= (-64, -23, -16).$$

The area of the base, $ABCE$, is

$$|(\bar{b} - \bar{a}) \times (\bar{c} - \bar{a})| = |(-45, 88, 59)| = \sqrt{45^2 + 88^2 + 59^2}$$
$$= \sqrt{13{,}250} = 115.108.$$

The length of the altitude to the base $ABCE$ is the quotient of the volume by the area of the base,

$$|\bar{p}| = 981/115.108 = 8.522.$$

8.10. THE EQUATIONS OF A LINE IN 3-SPACE.

The two distinct points

$$A = (a_1, a_2, a_3) \quad \text{and} \quad B = (b_1, b_2, b_3)$$

determine a line in 3-space. The vector

$$\overrightarrow{AB} = \bar{b} - \bar{a} = (b_1 - a_1, b_2 - a_2, b_3 - a_3)$$

is said to *lie on* the line AB. The direction of the line is the direction of the vector $(\bar{b} - \bar{a})$, the coordinates of which are the *direction numbers* of line AB.

If a point $P = (x, y, z)$ is to belong to line AB, then the vector $\overleftarrow{PA} = (\bar{p} - \bar{a})$ must have the same direction as the vector $\overrightarrow{AB} = (\bar{b} - \bar{a})$, so that

$$\overleftarrow{PA} = k\,\overrightarrow{AB}, \quad \text{or} \quad \overleftarrow{PA} \times \overrightarrow{AB} = 0.$$

The first equation gives us that

$$(x - a_1, y - a_2, z - a_3) = k(b_1 - a_1, b_2 - a_2, b_3 - a_3)$$
$$= (k(b_1 - a_1), k(b_2 - a_2), k(b_3 - a_3)).$$

When we use the first and third members and equate coordinates, we obtain

$$x - a_1 = k(b_1 - a_1), \qquad y - a_2 = k(b_2 - a_2), \qquad z - a_3 = k(b_3 - a_3),$$

from which we obtain, by simple algebraic methods, that

$$\frac{x - a_1}{b_1 - a_1} = \frac{y - a_2}{b_2 - a_2} = \frac{z - a_3}{b_3 - a_3} = k.$$

The equation, $\overleftarrow{PA} \times \overrightarrow{AB} = 0$ will lead to the same result.

For an example, we determine the equations of the line AB, where

$$A = (5, -3, 4) \qquad \text{and} \qquad B = (-6, 7, -2).$$

The direction numbers of line AB are

$$\overline{AB} = (\bar{b} - \bar{a}) = (-11, 10, -6).$$

If $P = (x, y, z)$ is to lie on line AB, then

$$\overleftarrow{PA} = \bar{p} - \bar{a} = (x - 5, y + 3, z - 4)$$

must have the same direction as AB. Consequently, the equations of line AB are

$$\frac{x - 5}{-11} = \frac{y + 3}{10} = \frac{z - 4}{-6}.$$

8.11. THE EQUATION OF A PLANE IN 3-SPACE. The three noncollinear points

$$A = (a_1, a_2, a_3), \qquad B = (b_1, b_2, b_3), \qquad C = (c_1, c_2, c_3)$$

determine a plane in 3-space. This plane, ABC, rests upon the terminal points of the three vectors

$$\bar{a} = (a_1, a_2, a_3), \qquad \bar{b} = (b_1, b_2, b_3), \qquad \bar{c} = (c_1, c_2, c_3).$$

The two vectors with common initial point at A,

$$\bar{b} - \bar{a} = (b_1 - a_1, b_2 - a_2, b_3 - a_3), \quad \text{and}$$
$$\bar{c} - \bar{a} = (c_1 - a_1, c_2 - a_2, c_3 - a_3),$$

lie on the plane *ABC*. The cross product vector

$$(\bar{b} - \bar{a}) \times (\bar{c} - \bar{a})$$

is perpendicular to plane *ABC*, and it is called the *normal vector* to the plane. In order that a point $P = (x, y, z)$ lie on plane *ABC*, the vector $(\bar{p} - \bar{a})$ must be perpendicular to the normal vector. Consequently, in order that *P* lie on the plane *ABC*, we must have

$$[(\bar{b} - \bar{a}) \times (\bar{c} - \bar{a})] \cdot (\bar{p} - \bar{a}) = 0,$$

which may be written in determinant form as follows:

$$[(\bar{p} - \bar{a}), (\bar{b} - \bar{a}), (\bar{c} - \bar{a})] = 0.$$

For an example, we calculate the equation of the plane which contains the three points

$$A = (3, -1, 6), \qquad B = (4, 2, -5), \qquad C = (-7, 4, 3).$$

The vectors

$$(\bar{b} - \bar{a}) = (1, 3, -11) \quad \text{and}$$
$$(\bar{c} - \bar{a}) = (-10, 5, -3)$$

lie on the plane *ABC*. The normal vector to this plane is

$$(\bar{b} - \bar{a}) \times (\bar{c} - \bar{a}) = (46, 113, 35).$$

The point $P = (x, y, z)$ must lie on the plane *ABC* if $(\bar{p} - \bar{a})$ is perpendicular to the normal vector; that is, if

$$(\bar{p} - \bar{a}) \cdot (46, 113, 35) = 46(x - 3) + 113(y + 1) + 35(z - 6) = 0.$$

8.12. THE DISTANCE FROM A POINT TO A LINE.   Given the points $A = (a_1, a_2, a_3)$ and $B = (b_1, b_2, b_3)$, which determine the line *AB*, and the point $C = (c_1, c_2, c_3)$, which does not lie on line *AB*. We are interested in finding the distance, *h*, from *C* to line *AB*.

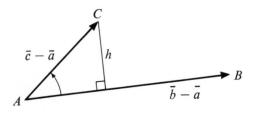

The vector $(\bar{b} - \bar{a})$ lies on line AB, and the vector $(\bar{c} - \bar{a})$ connects C to A. Since the distance h is one side of a right triangle of which $(\bar{c} - \bar{a})$ is the hypotenuse, it follows that

$$h = |\bar{c} - \bar{a}| \, |\sin A|.$$

But the length of the cross product vector gives us

$$|(\bar{b} - \bar{a}) \times (\bar{c} - \bar{a})| = |(\bar{b} - \bar{a})| \, |(\bar{c} - \bar{a})| \, |\sin A| = |(\bar{b} - \bar{a})|h,$$

from which we immediately obtain

$$h = \frac{|(\bar{b} - \bar{a}) \times (\bar{c} - \bar{a})|}{|(\bar{b} - \bar{a})|}.$$

For an example, we calculate the distance from the point $P = (-1, 2, 6)$ to the line determined by $A = (2, 3, -4)$ and $B = (8, 6, -8)$. We have that

$$(\bar{p} - \bar{a}) = (-3, -1, 10),$$

$$(\bar{b} - \bar{a}) = (6, 3, -4), \quad \text{and} \quad |(\bar{b} - \bar{a})| = \sqrt{61}.$$

Then,

$$|(\bar{p} - \bar{a}) \times (\bar{b} - \bar{a})| = |(-26, 48, -3)| = 7\sqrt{61},$$

and the distance, h, from P to line AB is

$$h = 7\sqrt{61}/\sqrt{61} = 7.$$

8.13. THE DISTANCE FROM A POINT TO A PLANE. Given the plane which passes through the point $A = (a_1, a_2, a_3)$ and has the normal vector $\bar{n} = (n_1, n_2, n_3)$. The normal vector, of course, may be computed, if we know two other points, B and C, which lie on the plane, as the vector product $\bar{n} = (\bar{b} - \bar{a}) \times (\bar{c} - \bar{a})$. We want to calculate the

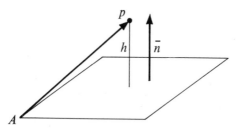

distance h from a point $P = (p_1, p_2, p_3)$ to the plane, where P does not lie on the plane.

The required distance, h, is equal to the length of the projection of the vector $(\bar{p} - \bar{a})$ upon the normal vector $\bar{n}$. This can be obtained from the dot product of $\bar{n}$ with $(\bar{c} - \bar{a})$, since

$$|\bar{n} \cdot (\bar{p} - \bar{a})| = |\bar{n}|h,$$

from which we obtain that

$$h = \frac{|\bar{n} \cdot (\bar{p} - \bar{a})|}{|\bar{n}|}.$$

For an example, we shall calculate the distance, h, from the point $P = (1, -5, 4)$ to the plane determined by the points

$$A = (3, -1, 6), \qquad B = (4, 2, -5), \qquad C = (-7, 4, 3).$$

The normal vector to plane ABC is

$$\begin{aligned}\bar{n} &= (\bar{b} - \bar{a}) \times (\bar{c} - \bar{a}) = (1, 3, -11) \times (-10, 5, -3) \\ &= (46, 113, 35).\end{aligned}$$

$$\begin{aligned}\bar{n} \cdot (\bar{p} - \bar{a}) &= (46, 113, 35) \cdot (-2, -4, -2) = -92 - 452 - 70 \\ &= -614.\end{aligned}$$

The length of the normal vector is

$$|\bar{n}| = \sqrt{46^2 + 113^3 + 35^2} = \sqrt{16{,}110} = 126.92.$$

Finally, we have

$$h = 614/126.92 = 4.837.$$

PROBLEM SET 8.1

1. Determine the length of each side, the cosine of each angle, and the area of triangle ABC, if
   a. $A = (1, 2, 0)$, $B = (-1, 4, 2)$, $C = (0, 2, 6)$;
   b. $A = (0, -1, 1)$, $B = (2, 0, 1)$, $C = (3, -2, 5)$;
   c. $A = (1, 1, 1)$, $B = (3, 4, 3)$, $C = (3, 0, 11)$.

2. Find the equations of the line through A and B, if
   a. $A = (16, -6, 4)$ and $B = (1, -2, 3)$;

    b. $A = (6, 3, -2)$ and $B = (7, 5, 3)$;

    c. $A = (22, -3, 2)$ and $B = (8, 3, 6)$.

3.   Find the equation of the plane determined by A, B, C, if

    a. $A = (5, -1, 4)$, $B = (1, 1, 8)$, $C = (4, 13, 5)$;

    b. $A = (7, 2, 1)$, $B = (2, 3, 2)$, $C = (-10, 2, 0)$;

    c. $A = (12, 3, 5)$, $B = (3, 4, 10)$, $C = (-6, 15, 5)$.

4.   Find the distance from P to line AB if

    a. $P = (1, -2, 6)$, $A = (3, 2, -4)$, $B = (8, -6, 8)$;

    b. $P = (5, -2, 4)$, $A = (2, 1, 8)$, $B = (5, 17, 4)$;

    c. $P = (6, -4, 10)$, $A = (5, 3, 4)$, $B = (13, 11, 12)$.

5.   Find the distance from P to the plane ABC

    a. $P = (-1, 1, 2)$, $A = (-3, 10, -9)$, $B = (2, -6, 6)$,
        $C = (1, 2, 3)$;

    b. $P = (5, -4, 1)$, $A = (2, -3, 2)$, $B = (8, 1, 1)$,
        $C = (1, 3, -1)$;

    c. $P = (4, -3, 3)$, $A = (-1, 7, -7)$, $B = (10, -5, 7)$,
        $C = (2, 5, 2)$.

6.   For each of the sets of four points in Problem 5 calculate the volume of the parallelopiped of which the given points are four of the vertices. Determine the remaining four points for each parallelopiped.

7.   Find the numerical value of the determinant $[\bar{a}, \bar{b}, \bar{c}]$ if

    a. $\bar{a} = (5, -16, 15)$, $\bar{b} = (4, -8, 12)$, $\bar{c} = (2, -9, 11)$;

    b. $\bar{a} = (3, -1, -1)$, $\bar{b} = (6, 4, -1)$, $\bar{c} = (-1, 6, -3)$;

    c. $\bar{a} = (6, -2, 4)$, $\bar{b} = (-2, 8, -1)$, $\bar{c} = (-5, 10, -10)$.

8.14. THE INTERSECTION OF THREE PLANES. We wish to find the point of intersection of the three planes

8.14.1.
$$P_1 = a_1x + a_2y + a_3z - d_1 = 0,$$
$$P_2 = b_1x + b_2y + b_3z - d_2 = 0,$$
$$P_3 = c_1x + c_2y + c_3z - d_3 = 0.$$

We make use of the concept of family, which was described in Section 7.6. The set of equations

8.14.2. $$k_1P_1 + k_2P_2 + k_3P_3 = 0,$$

where $k_1, k_2, k_3 \in K$ may not all be 0 simultaneously, represents the family of planes each member of which contains the point of intersection of all three planes. If we find the particular members of this family with equations of the form $x = r$, $y = s$, $z = t$, then we shall know that the point of intersection is (r, s, t).

We may form the matrix of detached coefficients and perform elementary row operations in the same way as was described in Section 7.7 until we obtain the matrix

$$\begin{pmatrix} 1 & 0 & 0 & r \\ 0 & 1 & 0 & s \\ 0 & 0 & 1 & t \end{pmatrix}.$$

Another interesting method makes use of determinants of order 2, and a pivot element. We detach the coefficients from the set of equations 8.14.1. Then we select the element a_1 as a *pivot* element for a pattern of determinants of order 2, as is demonstrated in the following array:

x	y	z	const.
a_1	a_2	a_3	d_1
b_1	b_2	b_3	d_2
c_1	c_2	c_3	d_3

$$A_2 = \begin{vmatrix} a_1 & a_2 \\ b_1 & b_2 \end{vmatrix}, \qquad A_3 = \begin{vmatrix} a_1 & a_3 \\ b_1 & b_3 \end{vmatrix},$$

$$D_1 = \begin{vmatrix} a_1 & d_1 \\ b_1 & d_2 \end{vmatrix}, \qquad B_2 = \begin{vmatrix} a_1 & a_2 \\ c_1 & c_2 \end{vmatrix},$$

$$B_3 = \begin{vmatrix} a_1 & a_3 \\ c_1 & c_3 \end{vmatrix}, \qquad D_2 = \begin{vmatrix} a_1 & d_1 \\ c_1 & d_3 \end{vmatrix},$$

Step 1.

	A_2	A_3	D_1
	B_2	B_3	D_2

Step 2. C_3 D_3

$$C_3 = \begin{vmatrix} A_2 & A_3 \\ D_2 & D_3 \end{vmatrix}, \quad \text{and} \quad D_3 = \begin{vmatrix} A_2 & D_1 \\ B_2 & D_2 \end{vmatrix}.$$

In the formation of determinants C_3 and D_3 the element A_2 is the pivot element.

In Step 2, the matrix $(C_3 \quad D_3)$ represents the detached coefficients of the equation

$$C_3z = D_3,$$

so, we have

$$z = D_3/C_3.$$

From $A_2y + A_3z = D_1$, we obtain

$$A_2y = -A_3D_3/C_3, \qquad y = -A_3D_3/A_2C_3.$$

From $a_1x + a_2y + a_3z = d_1$ (or from one of the other original equations, the choice of which should be in respect to whichever equation is easiest to use) we obtain

$$a_1x = -a_2y - a_3z + d_1 = a_2A_3D_3/A_2C_3 - a_3D_3/C_3 + d_1,$$

from which we have

$$x = a_2A_3D_3/a_1A_2C_3 - a_3D_3/a_1C_3 + d_1/a_1.$$

The point of intersection is the point (x, y, z) with the calculated values of x, y, and z just obtained.

For an example, we find the intersection of the three planes

$$2x + 3y + 6z = 1,$$
$$5x - 4y + z = 1,$$
$$3x + 5y + 7z = 2.$$

The array of detached coefficients, and the computation, with the element 2 in the first row and first column used for the pivot, is

2	3	6	1
5	-4	1	1
3	5	7	2
	-23	-28	-3
	1	-4	1
		120	-20

(1) $120z = -20$, $z = -1/6$;
(2) $y - 4(-1/6) = 1$, so $y = 1/3$;
(3) $2x + 3(1/3) + 6(-1/6) = 1$,
     so $2x = 1$, and $x = 1/2$.
(4) The point of intersection of the three planes is $(1/2, 1/3, -1/6)$.

PROBLEM SET 8.2

1. Find the point of intersection of the following planes and prove that these planes are mutually perpendicular.

$$2x + 3y - z = 3,$$
$$5x - 3y + z = 5,$$
$$y + 3z = 2.$$

2. Use (a) the method which uses elementary row operations on the matrix of detached coefficients, (b) the method using determinants of order 2, with the pivot element, to find the solution of the set of equations:

$$3x + y - z = 11,$$
$$x + 3y - z = 13,$$
$$x + y - 3z = 11.$$

3. The two methods used in Problem 2 may be extended to find the solution of similar intersection problems in higher dimensions than 3-space. Use each of these methods to find the point in 4-space in which the following four hyperplanes intersect:

$$2x - y + 3z - w = 9,$$
$$x - 4y + z = 11,$$
$$3x - 5z + 2w = 13,$$
$$8x + y + 4z - 2w = 30.$$

CHAPTER NINE

The Circular Functions

9.1. THE REFERENCE FRAME FOR THE CIRCULAR FUNCTIONS.   The reference frame for the circular functions, which will be established in the succeeding sections, consists of the cartesian plane with its x axis and y axis and the unit circle

$$\{(x, y) \; ; \; x^2 + y^2 = 1\},$$

each point of which is made to correspond to a certain set of real numbers.

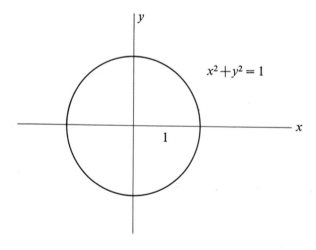

We construct a cartesian line with the same unit distance as that in the cartesian plane and consider this cartesian line to be a very fine, flexible, nonelastic filament. We fasten the point on the cartesian thread which corresponds to the real number 0 to the point $(1, 0)$ which lies on the unit circle and on the x axis. Then, we wind the positive half of the cartesian thread around the unit circle in the counterclockwise direction. This will establish a correspondence between the set of positive real numbers and the points on the unit circle, such that for every positive real number there is a point on the unit circle and every point on the unit circle corresponds to a certain set of real numbers. Next, the negative half of the cartesian thread is wrapped around the unit circle in the clockwise direction, thus establishing a correspondence between the points on the unit circle and the negative real numbers. For every negative real number there is a point on the unit circle and for every point on the unit circle there corresponds a certain set of negative real numbers.

Since the radius of the unit circle is 1, we know, from elementary geometry, that the circumference is 2π. The process described in the above paragraph establishes the following examples of the correspondence between selected points on the unit circle and subsets of K:

$(1, 0)$ corresponds to the set $\{2n\pi \; ; n \in I\}$;
$(0, 1)$ corresponds to the set $\{(4n + 1)\pi/2 \; ; n \in I\}$;
$(-1, 0)$ corresponds to the set $\{(2n + 1)\pi \; ; n \in I\}$;
$(0, -1)$ corresponds to the set $\{(4n - 1)\pi/2 \; ; n \in I\}$.

Thus, the reference frame for the circular functions establishes, in addition to the 1 : 1 correspondence between the set K_2 of ordered pairs of real numbers and the points in the cartesian plane, the correspondence between the set K of real numbers and the set of points on the unit circle. The point $T = (x, y)$ on the unit circle corresponds to a unique ordered pair of real numbers from the set

$$\{(x, y) \; ; x^2 + y^2 = 1\},$$

and this same point T corresponds to a certain subset of K. If the smallest positive real number which corresponds to T on the unit circle is t, then the point T corresponds to the set

$$\{t + 2n\pi \; ; n \in I\}.$$

9.2. THE COSINE OF t AND THE SINE OF t.   We select a point T on the unit circle and designate by t the smallest positive real number corresponding to T. Thus, T corresponds to the set

$$\{t + 2n\pi \; ; n \in I\}.$$

Since the coordinates of T depend upon the real number t, we designate, arbitrarily, the coordinates of $T = (x, y)$ to be

$$x = \cos t \quad \text{and} \quad y = \sin t,$$

where "$\cos t$" is read "cosine of t" and "$\sin t$" is read "sine of t." Thus, the coordinates of T may be written

$$T = (x, y) = (\cos t, \sin t).$$

In the following sketch of the reference frame for the circular functions, the abscissa, or first coordinate, of T is the distance repre-

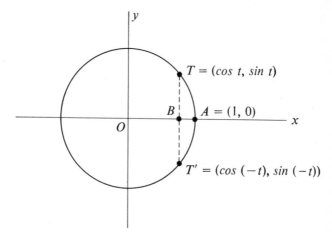

sented by the line segment OB, and the ordinate, or second coordinate, of T is the distance represented by the line segment BT, so

$$OB = \cos t \quad \text{and} \quad BT = \sin t.$$

Since T lies on the unit circle its coordinates must satisfy the equation of the unit circle, so we have that

$$\cos^2 t + \sin^2 t = 1.$$

Since T corresponds to the set $\{t + 2n\pi \; ; n \in I\}$, it follows that

$$\cos(t + 2n\pi) = \cos t, \quad \text{and}$$
$$\sin(t + 2n\pi) = \sin t.$$

In the sketch, the point T' on the unit circle corresponds to the real number $(-t)$. We see that

$$\cos(-t) = OB = \cos t, \quad \text{and}$$
$$\sin(-t) = BT' = -BT = -\sin t.$$

In the study of elementary functions we find that (i) if a function $f(t)$ has the property that, for every value of t, $f(-t) = f(t)$, then $f(t)$ is called an *even function;* and that (ii) if a function $f(t)$ has the property that, for every value of t, $f(-t) = -f(t)$, then $f(t)$ is called an *odd function*. Consequently, it follows that

$\cos t$ is an even function, and

$\sin t$ is an odd function.

We have been calling $\cos t$ and $\sin t$ functions without any justification that this is proper. If we consider the sets of ordered pairs defined as follows:

$$C = \{(t, \cos t) \,;\, t \in K\}, \quad \text{and,}$$
$$S = \{(t, \sin t) \,;\, t \in K\},$$

then, each of C and S is obviously a relation. Moreover, in C the ordered pairs $(t_1, \cos t_1)$ and $(t_2, \cos t_2)$ will be identical if $t_2 = t_1$, because this will require that $\cos t_2 = \cos t_1$. On the other hand, $\cos t_2 = \cos t_1$ does not imply that $t_2 = t_1$, since if $t_2 = t_1 + 2n\pi$, $n \in I$, then $\cos t_2 = \cos t_1$. As a consequence, no two distinct ordered pairs in C can have the same first element. The same remarks can be made about the ordered pairs in S. Both C and S are functions. It is customary, of course, to say in brief that each of $\cos t$ and $\sin t$ is a function.

The domain of each of $\cos t$ and $\sin t$ is the set K. The range for each of these functions is the set of real numbers included in the number interval $[-1, +1]$, since

$$-1 \le \cos t \le +1, \quad \text{and}$$
$$-1 \le \sin t \le +1.$$

Since $\cos t$ and $\sin t$ are functions which are closely related to the unit circle, we call them *circular functions*. There are six circular functions. The remaining four circular functions will be defined in the succeeding sections.

We sketch the graph of the function $c = \cos t$ in the cartesian plane containing the horizontal t axis and the vertical c axis. If we select the same unit as in the cartesian plane in which we defined $\cos t$ and $\sin t$, we may proceed as follows: at each real number t on the horizontal t axis, measure the corresponding distance OB in the vertical, or c direction. Thus, it will be possible to plot every point $(t, \cos t)$ which belongs to the function $\cos t$. The graph will appear as in the sketch below. It should be obvious that for each number interval on the t axis of length 2π the entire range $-1 \leq \cos t \leq +1$ is covered. Such a function is called a *periodic function*. The $\cos t$ is a periodic function with period of 2π.

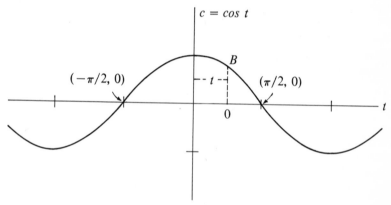

The graph of the function $s = \sin t$ is sketched in the same way. It looks like the following:

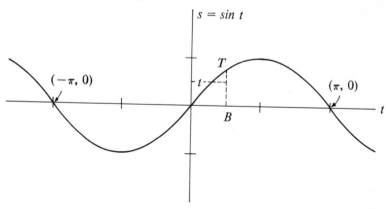

9.3. THE TANGENT OF t, THE COTANGENT OF t, THE SECANT OF t, AND THE COSECANT OF t. With the same reference frame consisting of the cartesian plane and the unit circle on which the real numbers are located, we construct a vertical line through the point $A = (1, 0)$. This vertical line is tangent to the unit circle at A. The radial line through O and T cuts the tangent line through A at the point C, and we are interested in determining the coordinates of the point C.

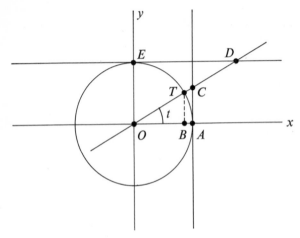

The abscissa of C is $OA = 1$. The ordinate of C is the length of the line segment AC. Triangle OBT is similar to triangle OAC, since each is a right triangle with the right angle at OBT and OAC, respectively, and the angle $BOT =$ angle AOC is common to both triangles. The equal ratios between corresponding sides of these similar triangles are

$$\frac{AC}{BT} = \frac{OA}{OB} = \frac{OC}{OT}.$$

Since $OA = OT = 1$, $OB = \cos t$, and $BT = \sin t$, we may write these ratios in the equivalent forms

$$\frac{AC}{\sin t} = \frac{1}{\cos t} = \frac{OC}{1}.$$

The first two members of this set of equal ratios gives us

$$AC = \frac{\sin t}{\cos t} = \tan t,$$

where "tan t" is read "tangent of t" and is the name given to the fraction

$$\tan t = \frac{\sin t}{\cos t}.$$

The coordinates of C are

$$C = (1, \tan t).$$

The second and third members of the set of equal ratios give

$$OC = \frac{1}{\cos t} = \sec t,$$

where "sec t" is read "secant of t." The sec t is the reciprocal of cos t.

We leave to the reader the proof that each of the relations $(t, \tan t)$ and $(t, \sec t)$ is a function. Since neither tan t nor sec t exists whenever $\cos t = 0$, the domain of each is the set

$$\{t \; ; t \in K, t \neq (2n + 1)\pi/2, n \in I\}.$$

The range of tan t is K. The range of sec t is the set

$$\{s \; ; s \in K, s \geq 1 \text{ or } s \leq -1\}.$$

Since, for every t, $\tan(-t) = -\tan t$, the tan t is an odd function. Since, for every t, $\sec(-t) = \sec t$, the sec t is an even function.

We leave to the reader to demonstrate that tan t is a periodic function with period of π and that sec t is a periodic function with period of 2π.

When we apply the well-known pythagorean theorem to triangle OAC we obtain

$$1 + \tan^2 t = \sec^2 t.$$

Now, we construct a horizontal line through the point $E = (0, 1)$. This horizontal line intersects the radial line determined by O and T at the point D. We are interested in finding the coordinates of D.

The ordinate of D is obviously 1. The abscissa of D is the length of the line segment ED. The triangles OBT and OED are similar since they are both right triangles with right angles at OBT and OED, respectively, and angle BOT is equal to angle ODE, since they are the alternate interior angles of the parallel lines OA and ED. The equal ratios from corresponding sides of the similar triangles OBT and OED are

$$\frac{ED}{OB} = \frac{OE}{BT} = \frac{OD}{OT}.$$

Since $OE = OT = 1$, $OB = \cos t$ and $BT = \sin t$, this set of equal ratios becomes

$$\frac{ED}{\cos t} = \frac{1}{\sin t} = \frac{OT}{1}.$$

From the first two members of this set of equal ratios we obtain

$$ED = \frac{\cos t}{\sin t} = \cot t,$$

where "cot t" is read "cotangent of t" and is the shortened form of the quotient of the $\cos t$ by the $\sin t$. The coordinates of D are $(\cot t, 1)$.

The second and third members in the set of equal ratios gives us

$$OT = \frac{1}{\sin t} = \csc t,$$

where "csc t" is read "cosecant of t."

Each of the relations $(t, \cot t)$ and $(t, \csc t)$ is a function. Each is an odd function. The $\cot t$ is periodic with period of π, and $\csc t$ is periodic with period of 2π.

Since neither $\cot t$ nor $\csc t$ exists whenever $\sin t = 0$, the domain of both functions is the set

$$\{t \; ; \, t \in K, t \neq n\pi, n \in I\}.$$

The range for $\cot t$ is K. The range for $\csc t$ is the set

$$\{s \; ; \, s \in K, s \geq +1 \text{ or } s \leq -1\}.$$

When we apply the pythagorean theorem to triangle OED we obtain

$$1 + \cot^2 t = \csc^2 t.$$

PROBLEM SET 9.1

1. Sketch the graph for each of the following functions:
    a. $(t, \tan t)$;
    b. $(t, \cot t)$;
    c. $(t, \sec t)$;
    d. $(t, \csc t)$.

2. By means of the reference frame described in Section 9.1 show that, for $n \in I$, each of the following is true:
    a. $\sin(t + 2n\pi) = \sin t$;
    b. $\cos(t + 2n\pi) = \cos t$;
    c. $\tan(t + n\pi) = \tan t$;
    d. $\cot(t + n\pi) = \cot t$;
    e. $\sec(t + 2n\pi) = \sec t$;
    f. $\csc(t + 2n\pi) = \csc t$.

9.4. THE CIRCULAR FUNCTIONS OF THE SUM, OR DIF-FERENCE, OF TWO REAL NUMBERS.   Given the real numbers t and u, with $t > u$. The point on the unit circle corresponding to t is $T = (\cos t, \sin t)$ and the point on the unit circle corresponding to u is $U = (\cos u, \sin u)$. We place the cartesian plane with the unit circle, as described in Section 9.1, into cartesian 3-space so that the coordinates of T and U are

$$T = (\cos t, \sin t, 0), \quad \text{and,}$$
$$U = (\cos u, \sin u, 0).$$

The vectors determined by the origin as initial point and the points T and U, respectively, as terminal points are

$$\overline{t} = (\cos t, \sin t, 0), \quad \text{and,}$$
$$\overline{u} = (\cos u, \sin u, 0).$$

Since the angle from $\overline{t}$ to $\overline{u}$ is a central angle of the unit circle it has the same measure, $(t - u)$, as the subtended arc. The length of each vector, $\overline{t}$ and $\overline{u}$, is 1.

The definition of the dot product gives us that

$$\overline{u} \cdot \overline{t} = 1 \cdot 1 \cdot \cos(t - u).$$

But the dot product is computed from the coordinates of the vectors to be

$$\bar{u}\cdot\bar{\imath} = \cos t \cos u + \sin t \sin u + 0.$$

Consequently, it follows that

$$\cos(t - u) = \cos t \cos u + \sin t \sin u.$$

Since $\cos t$ is an even function and $\sin t$ is an odd function we have

$$\cos(t + u) = \cos[t - (-u)] = \cos t \cos(-u) + \sin t \sin(-u)$$
$$= \cos t \cos u - \sin t \sin u.$$

The definition of the cross product gives us that

$$\bar{u} \times \bar{\imath} = 1\cdot 1\cdot\sin(t - u)\cdot\bar{n},$$

where $\bar{n}$ is a unit vector perpendicular to the plane containing $\bar{u}$ and $\bar{\imath}$. We know that the vector $k = (0, 0, 1)$ is perpendicular to the x-y plane. Since the determinant $[\bar{u}, \bar{\imath}, \bar{k}] = 1$ is positive it follows that $\bar{n} = \bar{k}$, and that

$$\bar{u} \times \bar{\imath} = \sin(t - u)\cdot\bar{k} = \sin(t - u)(0, 0, 1)$$
$$= (0, 0, \sin(t - u)).$$

The computation of the cross product gives us that

$$\bar{u} \times \bar{\imath} = (0, 0, \sin t \cos u - \cos t \sin u), \quad \text{so that}$$
$$(0, 0, \sin(t - u)) = (0, 0, \sin t \cos u - \cos t \sin u),$$

from which follows that

$$\sin(t - u) = \sin t \cos u - \cos t \sin u.$$

Again, from the fact that $\cos t$ is an even function and $\sin t$ is an odd function we have that

$$\sin(t + u) = \sin[t - (-u)] = \sin t \cos(-u) - \cos t \sin(-u)$$
$$= \sin t \cos u + \cos t \sin u.$$

Since $\tan(t - u) = \sin(t - u)/\cos(t - u)$, it follows that

$$\tan(t - u) = \frac{\sin(t - u)}{\cos(t - u)} = \frac{(\sin t \cos u - \cos t \sin u)/(\cos t \cos u)}{(\cos t \cos u + \sin t \sin u)/(\cos t \cos u)}$$

$$= \frac{\tan t - \tan u}{1 + \tan t \tan u}.$$

If we let $u = t$, we have $t + u = 2t$, and

$$\cos 2t = \cos^2 t - \sin^2 t, \quad \text{and}$$
$$\sin 2t = 2 \sin t \cos t.$$

PROBLEM SET 9.2

1. Prove each of the following equalities for circular functions:

a. $\tan(t + u) = \dfrac{\tan t + \tan u}{1 - \tan t \tan u}$;

b. $\cot(t - u) = \dfrac{\cot t \cot u + 1}{\cot t - \cot u}$;

c. $\tan 2t = \dfrac{2 \tan t}{1 - \tan^2 t}$;

d. $\cot 2t = \dfrac{\cot^2 t - 1}{2 \cot t}$.

Systems of Numeration

A,1. THE RECORDING OF NUMBERS, NUMERATION.

We remind the reader of the distinction between a number and the symbols used in recording the number in written form. We call the combination of symbols used to represent a number a *numeral*. If we have a systematic procedure for using a certain set of symbols to represent numbers we shall call the procedure a *system of numeration*.

Through the course of human history we find various methods for recording numbers. We recommend the development of numerals and of systems of numeration as a fascinating and profitable study. Certain vestiges of ancient methods for recording numbers are still in use. The modern scorekeeper, or data tabulator, makes use of the very ancient method of using slash marks to represent numbers. Thus,

> ⩘ represents five,
> ⩘ ||| represents eight, and
> ⩘ ⩘ || represents twelve.

We still find roman numerals in use, especially on façades of buildings to denote the year of construction of the building. Thus the roman numeral

> MCCMXLIX represents 1849,
> MCMXXIV represents 1924, and
> MCMLXVI represents 1966.

The history of the development of our present system of numeration based on ten symbols takes us from India and Arabia through the trade routes to Europe and the rest of the world. It is surprising to find that the full development of this system to its present form was attained only a relatively short time ago. Since our present system of numeration uses ten symbols it is called the *decimal system of numeration*.

It is claimed that we developed the decimal system of numeration simply because man has five fingers on each hand. If this assertion be true, then it would follow that if man had been created with six fingers on each hand, it would have been natural to develop a system of numeration based upon twelve symbols. Such a system will be discussed below, in Section A.3. It would also follow that if we had been created with only three fingers on each hand then our system of numeration would have six symbols. However, it is not really necessary to relate systems of numeration to counting on fingers. We shall, in succeeding sections, discuss various systems of numeration and establish the principles required for setting up a numeration system based on any selected number of symbols.

Not only is the study of various systems of numeration interesting but it also helps to provide the understanding of the algorithms for performing operations with numbers. In order to find the product, sum, or quotient of numbers in a strange system of numeration it is necessary that one understand fully the process required for obtaining the answer.

A.2. THE DECIMAL SYSTEM, BASE TEN. The system of numeration in universal use throughout the world is the *decimal system*, or *base ten system*. The decimal system makes use of the ten symbols

$$1, 2, 3, 4, 5, 6, 7, 8, 9, 0,$$

together with certain rules governing the place values of these symbols.

In writing the natural numbers in the decimal system, the symbols are written in a horizontal row with equal spacing between pairs of adjacent symbols. Each symbol has a definite place and its contribution to the number being represented depends on its place in the row of symbols. Starting from the right, the first place is called the "units place," the second place toward the left is called the "tens place," the third place is called the "hundreds place," the fourth place is called the "thousands place," and so on. The symbol in the tens place contributes ten times its numerical value, the symbol in the hundreds place contributes one hundred times its numerical value, the symbol in the thousands place contributes one thousand times its value, and so on. For example, the symbol 4367 in the decimal system means

$$4367 = 4000 + 300 + 60 + 7 = 4(10)^3 + 3(10)^2 + 6(10) + 7.$$

When we write successive natural numbers in the decimal system we find that the immediate successor of a numeral whose units digit is 9 has a change in the tens place; thus, the immediate successor

of 9 is 10,
of 19 is 20,
of 39 is 40,
of 99 is 100,
of 399 is 400,
of 1409 is 1410,
and so on.

We write one hundred in the decimal system as 100, one thousand is written 1000, ten thousand is written 10,000, and so on. Each of these numbers is a "power" of ten; that is,

$$100 = 10^2, \quad 1000 = 10^3, \quad 10,000 = 10^4, \quad \text{and so on.}$$

It follows that each natural number which is written in the decimal system of numeration can be expressed as the sum of multiples of powers of ten. For some examples,

$$783 = 7(10)^2 + 8(10) + 3,$$
$$4569 = 4(10)^3 + 5(10)^2 + 6(10) + 9,$$
$$3,476,485 = 3(10)^6 + 4(10)^5 + 7(10)^4 + 6(10)^3 + 4(10)^2 + 8(10) + 5.$$

When we write rational numbers, or approximations to irrational numbers, in the decimal system we make use of *decimal fractions*, in which we use the *decimal point*. First, the reader should recall the meaning of negative integers for exponents, so that

$$1/10 = 10^{-1}, \quad 1/100 = 10^{-2}, \quad 1/1000 = 10^{-3}, \quad 1/10,000 = 10^{-4}, \quad \text{etc.}$$

The decimal point is used to indicate a multiple of ten to a negative integral power, or the sum of multiples of negative integral powers of ten. Thus,

$$.1 = 1/10 = 10^{-1}, \quad .01 = 1/100 = 10^{-2},$$
$$.001 = 1/1000 = 10^{-3}, \quad .0001 = 1/10,000 = 10^{-4}, \quad \text{etc.}$$

Consequently, decimal fractions may be used to denote nonintegral numbers, as in the following examples:

$$.347 \; = 3(10)^{-1} + 4(10)^{-2} + 7(10)^{-3}$$
$$= \text{three hundred forty-seven thousandths,}$$
$$3.1416 = 3 + 1(10)^{-1} + 4(10)^{-2} + 1(10)^{-3} + 6(10)^{-4}$$
$$= \text{three and one thousand four hundred sixteen ten}$$
$$\text{thousandths,}$$
$$126.0032 = \text{one hundred twenty-six and thirty-two ten thousandths.}$$

We assume that the reader is familiar with the algorithms for adding, subtracting, multiplying and dividing with representations of numbers in the decimal system.

A.3. THE DOZEN SYSTEM, THE DUODECIMAL SYSTEM, BASE TWELVE.

The first system of numeration different from the decimal system that will claim our attention is based on the use of twelve symbols. We call this system the *dozen system*, or the *duodecimal system*, or the *base twelve system*. We shall employ the same familiar ten symbols that we use in the decimal system and two additional symbols. The new symbols are X, called *dec*, which is the base twelve numeral for the number ten, and Y, called *el*, which is the base twelve numeral for the number eleven. Consequently, the dozen system is based on the following twelve, or dozen, symbols:

$$1, 2, 3, 4, 5, 6, 7, 8, 9, X, Y, 0.$$

The symbol "10" is the numeral for twelve, or one dozen. When we write the number symbols in a horizontal row, the first symbol on the right end represents units, the second place toward the left represents the dozens place, the third symbol from the right represents the grosses (dozen dozens) place, and so on. Thus,

$$346 = \text{three gross four dozen and six} = 300 + 40 + 6.$$

Some attempts have been made to invent names for the place values in the dozen system. Thus, for example, two place numbers might use the suffix "do" so that "43" would read "fourdo and three," and the third place might use the suffix "gro," so that $3X7$ would read "threegro decdo and seven." Any such system might provide some fun but it would not really add to the understanding of the system.

One gross in the dozen system is written $100 = 10^2$, where the latter symbol means "one dozen squared." Similarly, a dozen gross is written $1000 = 10^3$, $10000 = 10^4$, and so on. It follows that every natural number can be written in the dozen system as the sum of multiples of integral powers of 10. For example,

$$78Y = 7(10)^2 + 8(10) + Y,$$
$$4X9Y = 4(10)^3 + X(10)^2 + 9(10) + Y.$$

In the multiplication table for the dozen system given here, the factor a is the numeral in the first column and the factor b is the numeral in the first row, and the product $a \times b$ occurs in the appropriate cell in the table.

b

x	1	2	3	4	5	6	7	8	9	X	Y	10
1	1	2	3	4	5	6	7	8	9	X	Y	10
2	2	4	6	8	X	10	12	14	16	18	$1X$	20
3	3	6	9	10	13	16	19	20	23	26	29	30
4	4	8	10	14	18	20	24	28	30	34	38	40
5	5	X	13	18	21	26	$2Y$	34	39	42	47	50
6	6	10	16	20	26	30	36	40	46	50	56	60
7	7	12	19	24	$2Y$	36	41	48	53	$5X$	65	70
8	8	14	20	28	34	40	48	54	60	68	74	80
9	9	16	23	30	39	46	53	60	69	76	83	90
X	X	18	26	34	42	50	$5X$	68	76	84	92	$X0$
Y	Y	$1X$	29	38	47	56	65	74	83	92	$X1$	$Y0$
10	10	20	30	40	50	60	70	80	90	$X0$	$Y0$	100

a

The Multiplication Table for the Dozen System

In the dozen system we may represent fractions by using the dozen point, or the duodecimal point. Thus, .1 is the symbol for 10^{-1}, or one dozenth (or one-twelfth), and .01 is the numeral for 10^{-2}, or one grossth. Some examples of numbers written as duodecimal fractions are

$$.7X9 = 7(10)^{-1} + X(10)^{-2} + 9(10)^{-3},$$
$$Y2.X94Y = Y(10) + 2 + X(10)^{-1} + 9(10)^{-2} + 4(10)^{-3} + Y(10)^{-4}.$$

Since the dozen system is simply another way for writing numbers in symbols and has the same general pattern as the decimal system, it is not surprising that the algorithms for performing operations are the same as are used in the decimal system. For example, we find the product of the duodecimal numbers 587×342 as follows:

```
          5 8 7
        × 3 4 2
        ────────
          Y 5 2
        1 X X 4
      1 5 1 9
    ──────────
    1 7 1 6 9 2
```

The division of the dozen number $5Y8X2$ by the dozen number 235 is done as follows:

```
              2 7 4 . 9 4 Y
      2 3 5 ⟌ 5 Y 8 X 2 . 0 0 0
              4 6 X
            ──────
            1 4 X X
            1 3 Y Y
          ────────
                X Y 2
                9 1 8
              ──────
                1 9 6 0
                1 8 6 9
              ────────
                  Y 3 0
                  9 1 8
                ──────
                  2 1 4 0
                  2 1 1 7
                ────────
                    2 5 = remainder.
```

A.4. INTERRELATIONS BETWEEN NUMERATION SYSTEMS. It is essential, of course, that it be clear which numeration system is being used. Unless it is stated otherwise it is assumed that the decimal system is being used. If one intends to use more than one numeration system in the same discourse we may use the notation in which a numeral is enclosed in parentheses with the subscript indicating the base of numeration. Thus,

$$(\text{numeral})_p$$

will mean that a number is being represented in the p-nary system, or base p system. Since the decimal system is the universal system we agree to write the base p in the decimal system, so

$$(\text{numeral})_{10}$$

means that the number is being written in the decimal system, and

$$(\text{numeral})_{12}$$

means that the number is being written in the dozen system.

This notation makes it possible to show the equivalent numerals for writing a number in different systems. For example, we have the following pairs of equivalent numerals for the same number:

$$(347)_{10} = (24Y)_{12},$$
$$(537)_{12} = (763)_{10},$$
$$(.325)_{12} = (3/12 + 2/144 + 5/1728)_{10} = (.26678\ldots)_{10}.$$

If we change the representation of a number from one system of numeration to another system we may compute the new numeral without a formal algorithm. For example, to find the base twelve numeral for $(347)_{10}$, we may, after one or more trials at each step, find that

$$(347)_{10} = (288 + 48 + 11)_{10} = (2(12)^2 + 4(12) + 11)_{10} = (24Y)_{12}.$$

Similarly, the equivalent decimal system numeral for $(537)_{12}$ may be determined as follows:

$$(537)_{12} = (5(10)^2 + 3(10) + 7)_{12} = (5(12)^2 + 3(12) + 7)_{10}$$
$$= (5(144) + 3(12) + 7)_{10} = (720 + 36 + 7)_{10} = (763)_{10}.$$

It may be somewhat faster, though less interesting, to use the following algorithm for changing the representation of a certain number from one numeration system to another, say from a base p system to a base q system. The algorithm requires successive division of (numeral)$_p$ by the representation of q in the p-nary system. For example, to find the base twelve numeral for $(347)_{10}$ by the algorithm we perform successive divisions of 347 by 12 and write the remainder at each step in the columns at the right:

		remainders	
		base 10	base 12
12	347		
	28	11	Y
	2	4	4
	0	2	2

The equivalent numeral in the dozen system is found by reading the column at the extreme right from the bottom upward, so

$$(347)_{10} = (24Y)_{12}.$$

For a second example, we find the dozen system equivalent of $(1966)_{10}$:

		remainders	
		base 10	base 12
12	1966		
	163	10	X
	13	7	7
	1	1	1
	0	1	1

Hence,

$$(1966)_{10} = (117X)_{12}.$$

When we use the same algorithm to find the numeral in the decimal system for the same number as (numeral)$_{12}$, we must divide successively by ten represented in the dozen system by the numeral X. For example, to find the decimal system equivalent for $(537)_{12}$ we divide successively by X and write the remainders in the columns at the right:

		base 12	base 10
X	537		
	64	3	3
	7	6	6
	0	7	7

Hence,

$$(537)_{12} = (763)_{10}.$$

For a second example, we find the equivalent decimal system numeral for $(17280)_{12}$:

X	17280	base 12	base 10
	1Y09	6	6
	238	1	1
	29	2	2
	3	3	3
	0	3	3

Hence, $(17280)_{12} = (33216)_{10}$. In this conversion, the second column of remainders is not necessary since each remainder must be less than ten, so that each remainder will have the same numeral in either system.

A.5. THE BINARY SYSTEM. The binary system of numeration has been very useful in the development of modern electronic computers. As its name implies, the *binary system*, or *base two system*, is constructed with only two symbols

$$1, 0.$$

The place value of each symbol in a numeral is a power of two. The numeral 10 represents two, 11 represents three, 100 represents the square of two, or four, and so on. The first twelve natural numbers represented in the binary system are:

1, 10, 11, 100, 101, 110, 111, 1000, 1001, 1010, 1011, 1100.

The tables for addition and multiplication in the binary system are

+	0	1
0	0	1
1	1	10

×	0	1
0	0	0
1	0	1

The following are equivalent numerals in the decimal system, dozen system and binary system

$$(27)_{10} = (23)_{12} = (11011)_2,$$
$$(587)_{10} = (40Y)_{12} = (1001001011)_2.$$

Fractions may be represented in the binary system by use of the *bimal point*. Thus, .1 means one-half, .01 means one fourth, .001 means one eighth, and so on. For examples of bimal numerals and the equivalents in the decimal system we have

$$(.1011)_2 = (1/2 + 0/4 + 1/8 + 1/16)_{10} = (.5 + .125 + .0625)_{10}$$
$$= (.6875)_{10} = (.83)_{12},$$
$$(101.01101)_2 = (5.40625)_{10} = (5.4X6)_{12}.$$

The algorithms for addition, multiplication, and division of numbers in the binary system are the same as we have used in the decimal system. For an example of multiplication we find the product of $(10111)_2$ and $(1101)_2$ as follows:

$$
\begin{array}{r}
1\,0\,1\,1\,1 \\
\times\, 1\,1\,0\,1 \\
\hline
1\,0\,1\,1\,1 \\
1\,0\,1\,1\,1 \\
1\,0\,1\,1\,1 \\
\hline
1\,0\,0\,1\,0\,1\,0\,1\,1
\end{array}
$$

For a long division example we have the division of $(1101101)_2$ by $(1011)_2$ as follows:

$$
\begin{array}{r}
1\,0\,0\,1\,.\,1\,1 \\
1\,0\,1\,1\,\overline{)\,1\,1\,0\,1\,1\,0\,1\,.\,0\,0} \\
1\,0\,1\,1 \\
\hline
1\,0\,1\,0\,1 \\
1\,0\,1\,1 \\
\hline
1\,0\,1\,0\,0 \\
1\,0\,1\,1 \\
\hline
1\,0\,0\,1\,0 \\
1\,0\,1\,1 \\
\hline
1\,1\,1 = \text{remainder}
\end{array}
$$

Crude electronic computers have been constructed based upon the fact that a row of lights can represent numbers in the binary system.

We may assign the value 1 to a light turned on and the value 0 to a light not turned on. In the following row of lights we represent a light turned on by $\otimes$ and one turned off by $\bigcirc$, so that

$$\otimes \quad \bigcirc \quad \bigcirc \quad \bigcirc \quad \otimes \quad \bigcirc \quad \otimes$$

represents the binary system numeral 1000101.

The algorithm for changing from one numeration system to another works in the same manner as explained in Section A.4. If we want to find the binary numeral for $(587)_{10}$ we proceed as indicated in the following:

2	587	remainders
	293	1
	146	1
	73	0
	36	1
	18	0
	9	0
	4	1
	2	0
	1	0
	0	1

Hence,

$$(587)_{10} = (1001001011)_2.$$

The same algorithm applied to changing $(Y5X)_{12}$ to the binary system is as follows:

2	Y 5 X	remainders
	5 8 Y	0
	2 X 5	1
	1 5 2	1
	8 7	0
	4 3	1
	2 1	1
	1 0	1
	6	0
	3	0
	1	1
	0	1

Hence,
$$(Y5X)_{12} = (11001110110)_2.$$

If we change from the binary system to the decimal system we must use the divisor $(1010)_2$, which represents ten, so that, for example, to find the representation in the decimal system for $(1011001)_2$ the calculation is as follows:

1010	1 0 1 1 0 0 1	base 2	base 10
	1 0 0 0	1 0 0 1	9 ↑
	0	1 0 0 0	8	

Hence,
$$(1011001)_2 = (89)_{10}.$$

We find the base twelve representation of the same number with the same algorithm:

1 1 0 0	1 0 1 1 0 0 1	base two	base twelve
	1 1 1	1 0 1	5 ↑
	0	1 1 1	7	

Hence,
$$(1011001)_2 = (75)_{12} = (89)_{10}.$$

A.6. OTHER SYSTEMS OF NUMERATION. Now that we have seen three systems of numeration it should be easy to construct other systems. We shall present the basis for starting the trinary, quadrinary, quinary, and base six systems. The reader may go further in each system and construct other numeration systems as he pleases.

A.6.1.   *The* trinary system *uses three symbols,*

$$1, 2, 0.$$

Each place value is a multiple of a power of three. The first twelve natural numbers written in the trinary system appear as follows:

$$1, 2, 10, 11, 12, 20, 21, 22, 100, 101, 102, 110.$$

A.6.2.   *The* quadrinary system *uses four symbols,*

$$1, 2, 3, 0.$$

Each place value is a multiple of a power of four. The first twelve natural numbers written in the quadrinary system appear as follows:

1, 2, 3, 10, 11, 12, 13, 20, 21, 22, 23, 30.

A.6.3. *The* quinary system *uses five symbols,*

1, 2, 3, 4, 0.

Each place value is a multiple of a power of five. The first twelve natural numbers written in the quinary system appear as follows:

1, 2, 3, 4, 10, 11, 12, 13, 14, 20, 21, 22.

A.6.4. *The* base six system *uses the six symbols,*

1, 2, 3, 4, 5, 0.

Each place value is a multiple of a power of six. The first twelve natural numbers written in the base six system appear as follows:

1, 2, 3, 4, 5, 10, 11, 12, 13, 14, 15, 20.

The algorithms are used in the same way in each numeration system. The only restriction is that in any given example all numbers must be written in the same numeration system. For an example, we find the product of $(234)_5$ and $(123)_5$ and then write each factor and the product in the decimal system.

$$
\begin{array}{r}
2\ 3\ 4 \\
\times\ 1\ 2\ 3 \\
\hline
1\ 3\ 1\ 2 \\
1\ 0\ 2\ 3 \\
2\ 3\ 4 \\
\hline
\end{array}
$$

$(4\ 0\ 4\ 4\ 2)_5$ product

We now change each factor to the decimal system, remembering to use ten, 20 in the quinary system, as the successive divisor.

(a) 20   234   base 5   base 10
          11   14   9 ↑ $(234)_5 = (69)_{10}.$
          0   11   6 |

(b) 20   123
          2   13   8 ↑ $(123)_5 = (38)_{10}.$
          0   3   3 |

(c)   For the product $(40442)_5$ we have

20	40442		
	2022	2	2
	101	2	2
	2	11	6
	0	2	2

$(40442)_5 = (2622)_{10}$

We shall let the reader verify that in base ten $69 \times 38 = 2622$.

APPENDIX EXERCISES

1. Make the tables for the operations of $+$ and $\times$ for (a) the trinary system, (b) the 7-nary system.
2. Convert $(2369)_{10}$ and $(875)_{10}$ to the dozen system. Find the product in the dozen system, then convert this product back to the decimal system and check whether it is correct.
3. Perform the following operations in the binary system, then convert the entire exercise into the decimal system:
   (a) $110101 + 1101 + 1001001 + 1111 =$,
   (b) $110110 \times 111000111 =$,
   (c) 111100011 divided by $1011 =$.
4. Given the decimal numbers $(3246)_{10}$ and $(5890)_{10}$.
   (a) Translate both numbers into the quinary system and find their product in the quinary system.
   (b) Translate the same numbers into the 7-nary system and find their sum and product in the 7-nary system.
5. The following are numbers in the decimal system. Convert each number to the binary system and then perform the indicated operations:
   (a) $19 + 168 - 57 =$,
   (b) $85 \times 193 =$.
6. Given the sets of lights in rows, so that they may represent numbers in the binary system according to the scheme described above. Perform the indicated operations, showing the results in a row of lights, and then convert the exercise to the binary system and the decimal system:
   (a) ⊗ ◯ ⊗ ◯ ◯ ⊗ + ⊗ ⊗ ◯ ⊗ ⊗ ◯ ⊗ ⊗ = ,
   (b) ⊗ ⊗ ◯ ◯ ⊗ ⊗ × ⊗ ◯ ⊗ ◯ = .

Index

193

86419